Boundary Element Analysis in Computational Fracture Mechanics

MECHANICS: COMPUTATIONAL MECHANICS

Editors: M. Stern and G.Æ. Oravas

T.A. Cruse, Boundary element analysis in computational fracture mechanics. 1988 ISBN 90-247-3614-5

Boundary Element Analysis in Computational Fracture Mechanics

T.A. Cruse

Department of Engineering Mechanics
Southwest Research Institute
San Antonio, TX, USA

KLUWER ACADEMIC PUBLISHERS
DORDRECHT / BOSTON / LONDON

Library of Congress Cataloging in Publication Data

Cruse, Thomas A.
Boundary element analysis in computational fracture mechanics
T.A. Cruse.
p. cm. -- (Mechanics--computational mechanics ; 1)
Includes bibliographies and index.

1. Fracture mechanics--Data processing. 2. Fracture mechanics--Mathematical models. 3. Boundary element methods. I. Title. II. Series.
TA409.C78 1988
620.1 126--dc19 88-9004
CIP

ISBN-13: 978-94-010-7118-5 e-ISBN-13: 978-94-009-1385-1
DOI: 10.1007/978-94-009-1385-1

Published by Kluwer Academic Publishers,
P.O. Box 17, 3300 AA Dordrecht, The Netherlands.

Kluwer Academic Publishers incorporates
the publishing programmes of
D. Reidel, Martinus Nijhoff, Dr W. Junk and MTP Press.

Sold and distributed in the U.S.A. and Canada
by Kluwer Academic Publishers,
101 Philip Drive, Norwell, MA 02061, U.S.A.

In all other countries, sold and distributed
by Kluwer Academic Publishers Group,
P.O. Box 322, 3300 AH Dordrecht, The Netherlands.

Softcover reprint of the hardcover 1st edition 1988

Dedication:

to Kathy, Michael, and Emily

PREFACE

The Boundary Integral Equation (BIE) method has occupied me to various degrees for the past twenty-two years. The attraction of BIE analysis has been its unique combination of mathematics and practical application. The BIE method is unforgiving in its requirement for mathematical care and its requirement for diligence in creating effective numerical algorithms. The BIE method has the ability to provide critical insight into the mathematics that underlie one of the most powerful and useful modeling approximations ever devised--elasticity. The method has even revealed important new insights into the nature of crack tip plastic strain distributions. I believe that BIE modeling of physical problems is one of the remaining opportunities for challenging and fruitful research by those willing to apply sound mathematical discipline coupled with physical insight and a desire to relate the two in new ways.

The monograph that follows is the summation of many of the successes of that twenty-two years, supported by the ideas and synergisms that come from working with individuals who share a common interest in engineering mathematics and their application. The focus of the monograph is on the application of BIE modeling to one of the most important of the solid mechanics disciplines--fracture mechanics. The monograph is not a treatise on fracture mechanics, as there are many others who are far more qualified than I to expound on that topic.

Rather, the monograph is a compilation of the various modeling strategies in which I have been involved, in advancing the use of BIE methods as a tool of computational mechanics, as applied to the analysis of cracked structures. The first chapter concerns some of the key historical developments, while Chapter 2 gives the briefest possible review of crucial fracture mechanics notions. Chapter 3 is of greatest importance to those not familiar with the BIE method. The purpose of this chapter is to define the basic mathematics of the BIE method in a manner consistent with my attitude towards mathematics--minimum reliance on theorems; maximum use of careful analysis. Chapter 4 is written to show that fracture mechanics is a poorly posed problem of elasticity, and that special mathematical treatments are required to deal with the posit that two surfaces can be in the same location--the usual definition of a crack.

Chapter 5 demonstrates one of the most powerful aspects of integral equation formulations, that the principal geometric feature of a problem (in this case the crack) may be included analytically in the formulation. Such an approach eliminates the need for modeling the crack. While limited to two dimensions, the resulting numerical technique is the best analysis method for two-dimensional cracks that is available. Chapter 6 demonstrates more recent work in extending this two-dimensional formulation to elastoplastic problems. Some unique insights into the plastic response around cracks have been established, and they warrant further study.

Chapters 7 and 8 are further, detailed applications of the BIE formulation to fracture mechanics modeling approaches that are still more in the research category. Thus, the algorithms reported are expected to undergo further development and improvement in the years ahead.

I owe much to many for the opportunity to have been actively involved in this area of research over the years. Each of these individuals shares one essential attribute--a commitment to excellence in research. Professor Frank J. Rizzo, who introduced me to the method and whose contributions underlie the BIE method; Professor J. L. Swedlow, who asked me to try modeling three-dimensional cracks and provided critical nurturing in my early research career; Dr. R. B. Wilson, who provided discipline and insight into the numerical issues; Dr. Jean-Claude Lachat, who saw the engineering potential of BIE analysis; and Professor John Watson, whose programming puts mine to shame. Additionally, I owe much to Professor C. L. Dym, who continues to challenge me to excellence of thought. The skilled preparation of this typescript is the result of the outstanding effort of Mrs. Maria Martinez; additionally, I wish to acknowledge the excellent support of my Administrative Assistant, Mrs. Cathy Dean, who brought the whole effort together, and the assistance of Ms. Deborah Stowitts and Ms. Maxine Hall.

Ultimately, when reflecting on the product of a significant part of one's research career, each must acknowledge the critical role of timing and circumstances in creating opportunities and revealing new directions. Over these years I have had the steadfast and foundational support of a person more talented than I--my wife, Kathy. And, when reflecting on the beauty of the mathematics that underlie all of this work, I acknowledge the One who created it all.

TABLE OF CONTENTS

LIST OF FIGURES

LIST OF TABLES

1.0 An Historical Perspective

1.1 Boundary Integral Equation Development

In the view of this author, the finite element and boundary element methods have a common root in the pioneering work of E. Betti (1872). Betti formulated a reciprocal work theorem, considering the work done by stresses in one solution state doing work on the strains of a distinct solution state. Such a "work" is work only in the mathematical sense and obviously has no physical counterpart.

However, structural engineers have long used the influence function approach formalized in the Castigliano theorems. These theorems have their basis in the Betti reciprocal work theorem and led into one of the major streams of early finite element developments. The notion of influence functions is also one of the easier ways to explain the BIE formulation to the non-mathematical engineer. Thus, a common origin between the boundary-integral equation and finite element methods may be said to exist.

The significant differences between the two methods, however, should not be discounted. Finite element methods approximate (in the usual formulations) the strain energy distribution through assumed displacement interpolations. Minimizing the energy functional with respect to the unknown displacement parameters produces an approximate equilibrium solution for the posed boundary conditions.

The boundary-integral equation formulation for the elasticity problem begins with a restatement of the equilibrium equations in integral form, using the reciprocal work theorem previously cited. Allowing one of the solution states to be one which exactly satisfies the equilibrium conditions in the infinite body has the result that the unknown solution also satisfies the equilibrium conditions, except possibly at the boundary. This analytical equilibrium solution is generally referred to as the fundamental solution for the linear, homogeneous differential equation governing the selected problem class. Formulation of the method is discussed in some detail in Chapter 3.

Somigliana (1885) is credited with the first use of the Betti reciprocal work theorem to derive the form of the integral solution identities based on these fundamental solutions. His work was in terms of the classical Laplace's equation for potential theory applications. It is most important to understand that the Somigliana reformulation of the potential theory problem corresponds, analytically, to the exact solution to Laplace's equation.

The fact that the Somigliana integral identities were fully equivalent to the exact, analytical solution led to their use by Fredholm (1905), for the development of existence and uniqueness proofs for the solutions. The Fredholm alternative theorems are quite elegant, but are quite limited in applicability. There is but one form of the Somigliana identity which results in the proper form of integral equation for application of these existence and uniqueness theorems.

A number of analytical solutions to potential theory problems have been obtained for very simple geometry and loading conditions. The Fredholm forms of the Somigliana identities for Laplace's equation have a particular relation to the force potential between mass distributions, from which the potential theory name developed. Some of the better sources of mathematical developments and applications of these methods are given in Lovitt (1950), Kellogg (1953), MacMillan (1958), and Courant and Hilbert (1962).

Application of the mathematics of integral formulations to elasticity problems was largely pursued by the Russian academics Muskhelishivili (1963) and by Kupradze (1965). In the former, the formulation depends on the Cauchy integrals of complex variable theory and is thereby limited to two-dimensional problems. In the latter, the Fredholm type of equations is derived such that existence and uniqueness questions may be posed and answered. The elasticity formulation results in integral equations which are said to be singular as compared to those for potential theory applications. Additional mathematical considerations, beyond those developed by Fredholm, had to be resolved, as discussed by Mikhlin (1965).

The complex variable formulation of the elasticity problem provided a significant tool for the solution of a variety of two-dimensional boundary value problems. The method requires a known or approximate geometric mapping function, capable of transforming the actual boundary shape into a unit circle, upon which the boundary conditions can be decomposed into Fourier components. The method is not applicable for problems with very complex shapes, or for mixed boundary conditions.

The digital computer became an engineering tool in the late 1950's and soon found application in the numerical solution of Fredholm type of integral equations (nonsingular). Early examples of this include the works of Friedman and Shaw (1962), Jaswon (1963), Banaugh and Goldsmith (1963), and Massonet (1965). The first, and certainly most critical, development of a viable computer solution to the singular elasticity formulation was that of Rizzo (1967). Following the paper of Rizzo, the numerical solution of integral equation formulations has been expanding at an ever increasing rate.

1.2 Boundary Formulations and Discretizations

As stated in the Section above, the integral equation formulation is one which exactly satisfies the governing differential equation. In the case of elasticity, this equation is the Navier equation of equilibrium, written in terms of the field displacements. The integral equations may be specifically developed for proofs of existence and uniqueness, or they may be developed for solving engineering problems. In general, one cannot have both.

In order to reduce the elasticity problem for traction boundary conditions to the proper form of the Fredholm equation for mathematical proofs, namely the Second Kind integral equation, it is necessary to replace the actual boundary conditions by non-physical surface density functions. The resulting integral equation has singularities in the density functions at any corner or edge of the geometry, as discussed by Cruse (1969).

Such singularities are inherent in all of the above-cited work except for that of Jaswon (1963) and Rizzo (1967). The radical departure from tradition started by Jaswon was crucial to the development of a viable method for the solution of engineering problems using integral equations. Jaswon's approach was to forego the ability to make elegant proofs of existence and uniqueness and make direct use of the full Somigliana integral identities. These identities are written directly in terms of the physical boundary conditions, regardless of the nature of the mixed boundary conditions being considered. Rizzo (1967) extended this direct approach to elasticity with great success. Thus, Cruse (1969) refers to the classical as the indirect (potential) method, and that of Jaswon and Rizzo as the direct (potential) method.

The term "direct potential method" was deemed by an editor to be a clumsy and obscure definition of a formulation. The direct formulation was then given the name boundary-integral equation (BIE) method by Swedlow and Cruse (1971). Somewhat after that time, other authors, notably Professors Banerjee and Brebbia, saw the need to relate the numerical capability of the method to the finite element method, and the name boundary element method (BEM) was coined. In this book, the formulation is referred to as the BIE, while the method of numerical implementation is referred to as the BEM.

The BIE formulation has been determined, by its nature, to be totally equivalent in elasticity to the equilibrium equation. Again, it is important to note the fundamental difference with finite element formulations which are in terms of the total strain energy or a virtual work representation. The numerical modeling of the BIE is in terms of approximations made to the boundary data--either specified or unknown.

The earliest numerical implementations of the BIE formulation used piecewise-constant models of the boundary data. In general, except for potential theory problems, this is not sufficiently accurate for any real problems. Later, Cruse (1974) implemented a piecewise-linear interpolation of both the surface geometry and boundary conditions. This resulted in a great improvement in accuracy, with some increase in the cost of computing the necessary terms in the algebraic system of equations.

Lachat and Watson (1976) were the first to implement the finite element-developed isoparametric representation for linear, quadratic, and cubic surface interpolatives. The method required numerical integration of the BIE terms and thereby required a very elaborate and painstaking set of algorithms, specially developed for the singular integral equations. The cost of computing the integrals increased significantly, but the modeling accuracy also increased. Since that time, little fundamentally new has been added to the formulation or numerical solution for elasticity problems.

The numerical form of the BIE is itself an equilibrium solution. That is, the internal variables fully satisfy the equations of equilibrium, due to the use of the fundamental solutions in the reciprocal work theorem. However, due to the boundary approximations, the numerical BIE is no longer the same equilibrium problem first posed. Rather, it is an equilibrium state that is trying to satisfy the sometimes contradictory states of piecewise variations of surface tractions and surface displacements. In fact, the equilibrium state includes various degrees of stress singularities at the interfaces between boundary elements, the strength of which depends on the types of elements and boundary data approximations being used. A proper understanding of this point is necessary before one can really see the significant and important differences between BIE and finite element methods. Further, a proper appreciation of this point is required in order to understand how little has really been accomplished in the way of a systematic investigation of numerical errors in BEM analysis.

1.3 Fracture Mechanics Problems

Fracture mechanics problems are among the most difficult to solve with reasonable numerical accuracy. The reason, of course, stems from the presence of a singular surface condition--two coplaner surfaces sharing a line connection (the crack front). This is a terribly ill-posed problem, and for some years the elasticians argued as to whether a solution should even be considered.

The reality of engineering structures and the success of a variety of numerical models, especially in two dimensions, led to a diminishing of theoretical criticisms. Nonetheless, the challenge for accuracy and generality of modeling capability was very great, especially in three dimensions. The BIE formulation has been found to be both strong and weak in its fracture mechanics modeling capabilities.

The weakness derives specifically from the ill-posed boundary of two coplaner surfaces. Special modeling approaches had to be developed to overcome the theoretical constraint, and these are discussed fully in Chapter 4. The strength is in the ability of the BIE to accurately represent steep solution gradients in elastic bodies. This ability comes from the fact that the BIE does not approximate the internal equilibrium solution or internal variables of the problem. This strength will also be discussed in Chapter 4.

Additionally, the BIE formulation has the ability to provide nearly analytical solutions and theoretical insight into the fracture mechanics problem. Again, this derives from the close tie between the method and the underlying equilibrium equations. Chapters 4 through 8 will make this point. The finite element method is, in fact, a more general tool than the BIE method in that energy methods can be used to approximate just about any physical phenomenon; however, the generality is achieved at the expense of significant approximation requirements.

Thus, in the end, the analyst must make the choice. The remaining chapters seek to provide analysts with the necessary support to understand the BIE method and to evaluate its applicability to a given problem class.

References

(1872) E. Betti, Teoria dell Elasticita, Il Nuovo Ciemento, t. 7-10.

(1885) C. Somigliana, Sopra l'equilibrio di us corpo elastico isotropo, Il Nuovo Ciemento, tome 17-19.

(1905) I. Fredholm, Solution d'un problem fundamental de la theorie de l'elasticite, Arkiv for Matematik, Astronomic och Fvsik, 2, 1-8.

(1950) W.V. Lovitt, Linear Integral Equations, Dover.

(1953) O.D. Kellogg, Foundations of Potential Theory, Dover.

(1958) W.D. MacMillan, The Theory of the Potential, Dover.

(1962) R. Courant and D. Hilbert, Methods of Mathematical Physics, Vol. II, Interscience.

(1962) M.B. Friedman and R. Shaw, Diffraction of Pulse by Cylindrical Obstacles of Arbitrary Cross Section, Journal of Applied Mechanics, Transactions of the ASME, 40-46.

(1963) R.P. Banaugh and W. Goldsmith, Diffraction of Steady Elastic Waves by Surfaces of Arbitrary Shape, Journal of the Acoustical Society of America, 35, 1590-1601.

(1963) M.A. Jaswon, Integral Equation Methods in Potential Theory, I, Proceedings of Royal Society (A), 275, 23-32.

(1963) N.I. Muskhelishivili, Some Basic Problems of the Mathematical Theory of Elasticity, Noordhoff.

(1965) V.D. Kupradze, Potential Methods in the Theory of Elasticity, D. Davey.

(1965) C.E. Massonet, Numerical Use of Integral Procedures, in Stress Analysis, (ed. O. C. Zienkiewicz and G. S. Hollister), Wiley.

(1965) S.G. Mikhlin, Multidimensional Singular Integrals and Integral Equations, Pergamon Press.

(1967) F.J. Rizzo, An Integral Equation Approach to Boundary Value Problems of Classical Elastostatics, Quarterly of Applied Mathematics, 25, 83-95.

(1969) T.A. Cruse, Numerical Solutions in Three Dimensional Elastostatics, International Journal of Solids & Structures, 5, 1259-1274.

(1971) J.L. Swedlow and T.A. Cruse, Formulation of Boundary Integral Equations for Three Dimensional Elastoplastic Flow, International Journal of Solids & Structures, 7, 1673-1683.

(1974) T.A. Cruse, An Improved Boundary-Integral Equation Method for Three Dimensional Elastic Stress Analysis, Computers & Structures, 4, 741-754.

(1976) J.C. Lachat and J.O. Watson, Effective Numerical Treatment of Boundary-Integral Equations, International Journal of Numerical Methods in Engineering, 10, 991-1005.

2.0 Fracture Mechanics

2.1 Introduction

The present monograph is concerned principally with boundary-integral equation methods, especially on their application to fracture mechanics modeling. The literature on fracture mechanics is quite extensive. For a comprehensive treatment of fundamentals, the reader is directed to Broek (1986), and for advanced discussions, to Kanninen and Popelar (1986).

The current Chapter is intended to define some of the basic concepts and terminologies of fracture mechanics as needed to support later discussions. Thus the review will define basic terms and cite principal mathematical results in two and three dimensions. For a review of non-BEM fracture mechanics modeling approaches (these are generally collocation and finite element results), the reader is directed to the aforementioned references and to their extensive bibliographies. Further, the literature on BEM treatments of various fracture mechanics problems is also extensive. The current monograph is not intended to be a review of these results; for that, the reader is directed to the article by Cruse (1987).

2.2 Some Definitions

In general, we will be concerned with the analysis of regular regions containing internal or edge surfaces across which the displacements are discontinuous. These cracks are usually flat, but may grow along curved paths or surfaces. In the current treatment we will be concerned with flat cracks.

Figure 2.1 illustrates the essential crack geometry features. In two-dimensional (plane strain) problems the crack will extend indefinitely in a direction normal to the plane of the figure. In three-dimensional problems, the crack front will form a curve in the $y=0$ plane (again, for the flat crack).

The crack-tip displacement field is decomposed into simple vector components according to Fig. 2.2. In the left-hand subfigure the crack deformation is symmetric with respect to the plane of the crack and is called Mode I response. The middle subfigure shows in-plane shear response

(γ_{xy}) and is called Mode II response. The antiplane shear on the right is called Mode III response. Two-dimensional problems generally consist of Mode I and Mode II deformations, while three-dimensional problems may involve all three modes.

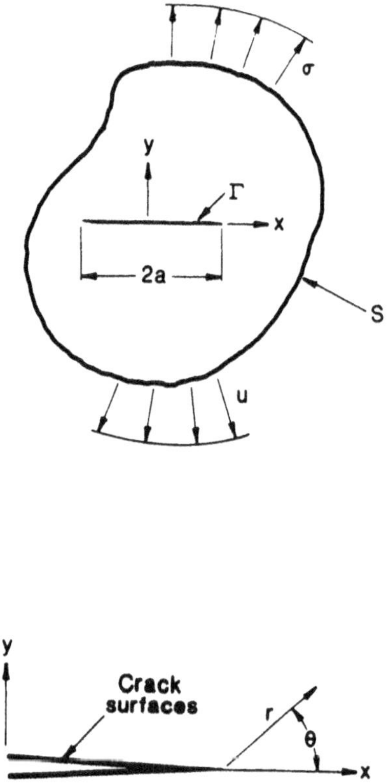

Figure 2.1. Crack geometry definition

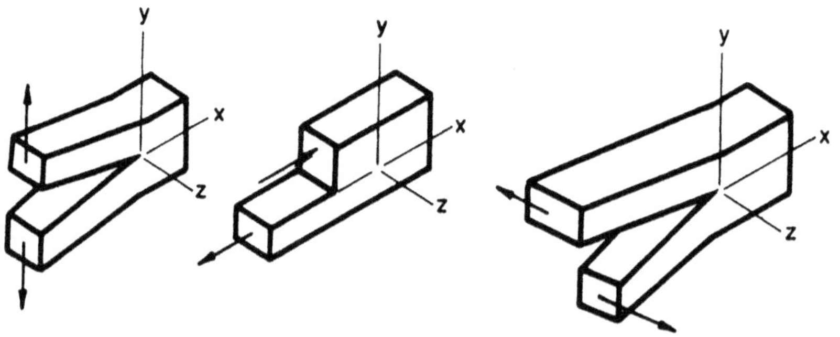

Figure 2.2. Three modes of crack tip motion

Three-dimensional cracks often are modelled as having elliptical shapes. Figure 2.3 defines the principal geometric terms for buried elliptical cracks. When the elliptical crack is at a free surface of the body, it is generally the case that the free surface is taken to be the plane z=0. Then, the total surface length of the crack is taken as 2c instead of 2a in Fig. 2.1. The crack depth is then given by the size parameter a.

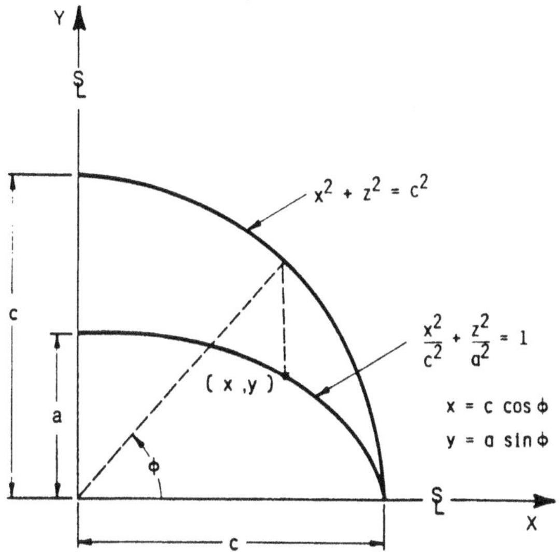

Figure 2.3. Geometry terms for elliptical crack solution

2.3 Some Fundamental Results

The earliest effort to characterize the strength of brittle materials in a manner that recognized the size of the defect or crack was that of Griffith (1920). In his analysis, the basic energy balance for a virtual crack length extension was first derived in terms of the elastic stored energy per unit thickness, U

$$G = -\partial U/\partial a \qquad (2.1)$$

where a is the half crack length, δa is the crack length change, and G is the strain energy release rate. Equating the release in stored energy to the increase in crack surface area energy (material property), for the Inglis (1913) plane strain problem of a central slit in tension, Griffith obtained the following basic relation

$$G = (1-\nu^2)\pi\sigma^2 a/E \qquad (2.2)$$

which applies up to the point of crack growth instability. In (2.2), σ is the applied far-field stress and E, ν are the usual material constants.

Irwin (1957) considered the state of stress ahead of the crack under load and recognized that the Griffith energy release rate could be described in terms of a parameter which characterized the state of stress or strain in the region around the crack tip. This parameter K, the stress intensity factor, is related to the energy release rate by the following

$$G = \begin{cases} K^2/E \text{ (plane stress)} \\ (1-\nu^2)\, K^2/E \text{ (plane strain)} \end{cases} \tag{2.3}$$

The key feature recognized by Irwin which became the foundation of linear elastic fracture mechanics (LEFM) is the fact that, for limited plasticity, the stress intensity factor (SIF) characterizes the state of material loading over a volume of sufficient size that the fracture strength of many engineering materials can be given in terms of the critical (maximum) stress intensity factor, K_{IC}.

Williams (1952) treated the general class of problems associated with a vertex in the material under load for various types of boundary conditions on the edges of the vertex as an eigenvalue problem. He deduced that the transcendental equation

$$\sin z = \pm\, [(\sin \alpha)/\alpha] z; \qquad z = \lambda\alpha \tag{2.4}$$

gives the power λ of the inverse singularity of stress or strain at the vertex tip for the case of stress-free edges near the tip. From (2.4), the stress singularity at the tip of the vertex is given by $r^{\lambda-1}$; α is the total included material angle at the vertex. The crack-tip stress solution for the case of $\alpha = 2\pi$ for remote loading is given by

$$\sigma = a_0 r^{-1/2} + a_1 + a_2 r^{+1/2} + \cdots \tag{2.5}$$

The leading term in (2.5) is seen to be singular at the crack tip, and thus for some region near the crack tip, dominates the near field stress and strain field. The coefficient of this leading term in Williams' solution is proportional to the SIF defined by Irwin. Using the current SIF definition, the leading term for the two-dimensional stress and displacement field near the crack tip (crack along x-axis) is given for plane strain by

$$\left.\begin{aligned}\sigma_x &= \frac{K_I}{(2\pi r)^{1/2}} \cos\frac{\theta}{2} \left(1 - \sin\frac{\theta}{2}\sin\frac{3\theta}{2}\right)\\ \sigma_y &= \frac{K_I}{(2\pi r)^{1/2}} \cos\frac{\theta}{2} \left(1 + \sin\frac{\theta}{2}\sin\frac{3\theta}{2}\right)\\ \tau_{xy} &= \frac{K_I}{(2\pi r)^{1/2}} \sin\frac{\theta}{2}\cos\frac{\theta}{2}\cos\frac{3\theta}{2}\end{aligned}\right\} \tag{2.6}$$

$$\left.\begin{aligned}u &= \frac{K_I}{\mu}(r/2\pi)^{1/2}\cos\frac{\theta}{2}\left(1 - 2\nu + \sin^2\frac{\theta}{2}\right)\\ v &= \frac{K_I}{\mu}(r/2\pi)^{1/2}\sin\frac{\theta}{2}\left(2 - 2\nu - \cos^2\frac{\theta}{2}\right)\end{aligned}\right\} \tag{2.7}$$

$$\left.\begin{aligned}\sigma_x &= \frac{-K_{II}}{(2\pi r)^{1/2}} \sin\frac{\theta}{2}\left(2 + \cos\frac{\theta}{2}\cos\frac{3\theta}{2}\right)\\ \sigma_y &= \frac{K_{II}}{(2\pi r)^{1/2}} \cos\frac{\theta}{2}\sin\frac{\theta}{2}\cos\frac{3\theta}{2}\\ \tau_{xy} &= \frac{K_{II}}{(2\pi r)^{1/2}} \cos\frac{\theta}{2}\left(1 - \sin\frac{\theta}{2}\sin\frac{3\theta}{2}\right)\end{aligned}\right\} \tag{2.8}$$

$$\left.\begin{aligned}u &= \frac{K_{II}}{\mu}(r/2\pi)^{1/2}\sin\frac{\theta}{2}\left(2 - 2\nu + \cos^2\frac{\theta}{2}\right)\\ v &= \frac{K_{II}}{\mu}(r/2\pi)^{1/2}\cos\frac{\theta}{2}\left(1 - 2\nu + \sin^2\frac{\theta}{2}\right)\end{aligned}\right\} \tag{2.9}$$

$$\left.\begin{aligned}\tau_{xz} &= -\frac{K_{III}}{(2\pi r)^{1/2}} \sin\frac{\theta}{2}\\ \tau_{yz} &= \frac{K_{III}}{(2\pi r)^{1/2}} \cos\frac{\theta}{2}\end{aligned}\right\} \tag{2.10}$$

$$w = \frac{K_{III}}{\mu}(2r/\pi)^{1/2}\sin\frac{\theta}{2} \tag{2.11}$$

In eqs. (2.6)-(2.11), μ is the shear modulus; θ is the angle from the plane of crack prolongation given in Fig. 2.1. The subscripts on the SIF terms in these equations refer to the three, kinematically independent modes of crack-tip motion as shown in Fig. 2.2.

The above results apply to the case of isotropic material behavior. While the essential singular behavior of the crack tip stresses and strains is retained in the rectilinearly anisotropic case, the leading term results in eqs. (2.6)-(2.11) are somewhat more complicated. Let β_{ij} describe the full (2D) material compliance matrix (up to 21 constants in the general 3D case). The governing differential equations for planar problems with elastic mid-plane symmetry are integrable along the complex directions given by the roots of the following characteristic equation (Sih and Liebowitz (1968))

$$\beta_{11}\mu^4 - 2\beta_{16}\mu^3 + (2\beta_{12} + \beta_{66})\mu^2 - 2\beta_{26}\mu + \beta_{22} = 0 \qquad (2.12)$$

The roots of (2.12) given in complex conjugates $\mu_k, \bar{\mu}_k$ are dimensionless quantities. The energy release rate is given for mode I, II loading as

$$G = -\frac{K_I^2}{2\pi}\beta_{22}\,\mathrm{Im}\left\{\frac{(\mu_1 + \mu_2)}{\mu_1\mu_2}\right\} + \frac{K_{II}^2}{2\pi}\beta_{11}\mathrm{Im}\,\{\mu_1 + \mu_2\} \qquad (2.13)$$

where Im{ } denotes the imaginary part. The crack tip stresses are given by

$$\left.\begin{aligned}
\sigma_x &= \frac{K_I}{\sqrt{2\pi r}}\,\mathrm{Re}\left[\frac{\mu_1\mu_2}{\mu_1-\mu_2}\left(\frac{\mu_2}{d_2} - \frac{\mu_1}{d_1}\right)\right] + \frac{K_{II}}{\sqrt{2\pi r}}\,\mathrm{Re}\left[\frac{1}{\mu_1-\mu_2}\left(\frac{\mu_2^2}{d_2} - \frac{\mu_1^2}{d_1}\right)\right]\\
\tau_{xy} &= \frac{K_I}{\sqrt{2\pi r}}\,\mathrm{Re}\left[\frac{\mu_1\mu_2}{\mu_1-\mu_2}\left(\frac{1}{d_1} - \frac{1}{d_2}\right)\right] + \frac{K_{II}}{\sqrt{2\pi r}}\,\mathrm{Re}\left[\frac{1}{\mu_1-\mu_2}\left(\frac{\mu_1}{d_1} - \frac{\mu_2}{d_2}\right)\right]\\
\sigma_y &= \frac{K_I}{\sqrt{2\pi r}}\,\mathrm{Re}\left[\frac{1}{\mu_1-\mu_2}\left(\frac{\mu_1}{d_2} - \frac{\mu_2}{d_1}\right)\right] + \frac{K_{II}}{\sqrt{2\pi r}}\,\mathrm{Re}\left[\frac{1}{\mu_1-\mu_2}\left(\frac{1}{d_2} - \frac{1}{d_1}\right)\right]
\end{aligned}\right\} \qquad (2.14)$$

where

$$d_j = (\cos\theta + \mu_j\sin\theta)^{1/2} \qquad (2.15)$$

2.4 Some Stress Intensity Factors

The SIF terms in eqs. (2.6)-(2.11) for two-dimensional results have been widely published in a number of handbooks. Other results, for cases not included in the handbooks, are developed by numerical stress analysis of the cracked geometry and loading conditions. The use of BIE methods for such problems is well developed and will be discussed in detail in the next chapters.

Three-dimensional SIF results are not as widely available as for two-dimensional problems. The earliest, classical result is by Green and Sneddon (1950) for the case of the buried, elliptical plan-form crack. The solution, in terms of the quantities defined in Fig. 2.3, is given by

$$K_I = \frac{\sigma\sqrt{\pi a}}{E(k)}\{\sin^2\phi + a^2/b^2\cos^2\phi\}^{1/4}(a<b)$$

$$\text{with } E(k) = \int_0^{\pi/2} \sqrt{1-k^2\sin^2\phi}\, d\phi = \sqrt{Q} \qquad (2.16)$$

$$\text{and } k^2 = 1 - a^2/b^2$$

The elliptic integral, in eq. (2.16), is the square root of the shape factor (Q) for elliptical cracks and is often used as a normalizing parameter for surface crack results. An extensive set of finite element numerical results for surface cracks in plates under tension and bending was curve fit by Newman and Raju (1981). Their results are reproduced below

$$K_1 = (S_t + HS_b)\sqrt{\pi \frac{a}{Q}}\, F(\frac{a}{t},\frac{a}{c},\frac{c}{b},\phi)$$

$$Q = 1 + 1.464\,(\frac{a}{c})^{1.65} \quad (\frac{a}{c} \leq 1).$$

$$F = [M_1 + M_2(\frac{a}{t})^2 + M_3(\frac{a}{t})^4]f_\phi g f_w$$

$$M_1 = 1.13 - 0.09(\frac{a}{c})$$

$$M_2 = -0.54 + \frac{0.89}{0.2 + (a/c)}$$

$$M_3 = 0.5 - \frac{1.0}{0.65 + (a/c)} + 14(1.0 - \frac{a}{c})^{24}$$

$$\left.\begin{aligned}
g &= 1 + [0.1 + 0.35(\tfrac{a}{t})^2](1 - \sin\phi)^2 \\
f_\phi &= [(\tfrac{a}{c})^2 \cos^2\phi + \sin^2\phi)]^{1/4} \\
f_w &= [\sec(\tfrac{\pi c}{2b}\sqrt{\tfrac{a}{t}})]^{1/2} \\
H &= H_1 + (H_2 - H_1)\sin^p\phi \\
p &= 0.2 + \tfrac{a}{c} + 0.6\,\tfrac{a}{t} \\
H_1 &= 1 - 0.34\,\tfrac{a}{t} - 0.11\,\tfrac{a}{c}\,(\tfrac{a}{t}) \\
H_2 &= 1 + G_1(\tfrac{a}{t}) + G_2(\tfrac{a}{t})^2. \\
G_1 &= -\ 1.22 - 0.12(\tfrac{a}{c}) \\
G_2 &= 0.55 - 1.05(\tfrac{a}{c})^{0.75} + 0.47(\tfrac{a}{c})^{1.5}
\end{aligned}\right\} \qquad (2.17)$$

In eq. (2.17), t is the section thickness, a the crack depth, c the crack half surface length, ϕ the elliptic angle (Fig. 2.3), b the plate half-width, Q the shape factor (1.16), H = E, or $E/(1-\nu^2)$, and S_t, S_b the remote applied tensile and bending stresses. These results generally agree, for shallow cracks in tension, with the earlier BIE results of Cruse and Meyers, (1977), but are more comprehensive and are of great engineering utility.

Extensive collections of stress intensity factors have been published in handbook form by Tada (1973) and Rooke and Cartwright (1976). Most of these solutions are for two-dimensional problems, while Newman and Raju (1981) should be used for surface cracks. Very powerful methods for constructing K-solutions from the uncracked stress fields have become popular, including the weight function method as summarized by Besuner (1977), and the influence function method, also called the Green's function method by Cartwright and Rooke (1979). In the latter approach, K-solutions for point loads on the two-dimensional crack surface are superimposed to match the uncracked stresses.

References

(1913) C.E. Inglis, Stresses in a Plate Due to the Presence of Cracks and Sharp Corners, Proceedings of the Institute of Naval Architects, 60, 219-230.

(1920) A.A. Griffith, The Phenomena of Rupture and Flow in Solids, Philosophical Transactions of the Royal Society, A, 221, 163-198.

(1950) A.E. Green and I.N. Sneddon, The Stress Distribution in the Neighborhood of a Flat Elliptical Crack in an Elastic Solid, Proceedings of the Cambridge Philosophical Society, 46, 159-163.

(1952) M.L. Williams, Stress Singularities Resulting from Various Boundary Conditions in Angular Corners of Plates in Extension, Journal of Applied Mechanics, 19, 526-528.

(1957) G.R. Irwin, Analysis of Stresses and Strains Near the End of a Crack Traversing a Plate, Journal of Applied Mechanics, 361-364.

(1968) G.C. Sih and H. Liebowitz, Fracture: An Advanced Treatise, Vol. 2, Academic Press, New York.

(1973) H. Tada, The Stress Analysis of Cracks Handbook, Del Research Corporation, Hellertown, PA.

(1976) D.P. Rooke and D.J. Cartwright, Compendium of Stress Intensity Factors, Her Majesty's Stationary Office, London.

(1977) P.M. Besuner, The Influence Function Method for Fracture Mechanics and Residual Fatigue Life Analysis of Cracked Components Under Complex Stress Fields, Nuclear Engineering and Design, 43, 1.

(1977) T.A. Cruse and G.J. Meyers, Three-Dimensional Fracture Mechanics Analysis, Journal of the Structural Division, American Society of Civil Engineers, 103, 309-320.

(1979) D.J. Cartwright and D.P. Rooke, Green's Functions in Fracture Mechanics: Current Status, Future Prospects, R.A. Smith, Ed., Pergamon Press, Oxford, 91-123.

(1981) J.C. Newman and I.S. Raju, An Empirical Stress Intensity Factor Equation for the Surface Crack, Engineering Fracture Mechanics, 15, 185-192.

(1985) M.F. Kanninen and C.H. Popelar, Advanced Fracture Mechanics, Oxford University Press, New York.

(1986) D. Broek, Elementary engineering fracture mechanics, Martinus Nijhoff Publishers, Dordrecht, The Netherlands.

(1987) T.A. Cruse, Fracture Mechanics, in Boundary Element Methods in Mechanics, (ed. D.E. Beskos), Elsevier Science Publishers B.V., 334-365.

3.0 Boundary-Integral Equation Formulation and Solution

3.1 Introduction

The basis of any boundary integral equation (BIE) formulation is rooted in classical elasticity theory as the reciprocal work theorem of Betti (1872) and the application of elastic potentials to satisfy equilibrium by Somigliana (1885). Much of the literature in the past ten years of BIE formulations has made use of the method of weighted residuals. While simple to apply, especially by those who have a finite element background, the method of weighted residuals masks much of the essentials to a full understanding of the strengths and weaknesses of the BIE method. We shall thus take the high road of the classical approach to the BIE method in what follows.

Much of the older (greater than twenty years) literature on integral equation formulations in mechanics ties back to classical work on gravity and electrostatic potentials. In fact, the BIE formulation for scalar problems such as steady-state heat transfer or potential flow use exactly the same integral equations as developed for gravity and electrostatic potentials. Thus, a familiarity in potential theory such as the books by Kellogg (1953) and Kupradze (1965) is essential for a full understanding of the BIE literature.

3.2 Governing Equations of Elasticity in Two and Three Dimensions

The current discussion will treat isotropic elastic continua, deferring to the next section the development for anisotropic materials. Equilibrium in the presence of body forces (per unit volume) X_i is given by the gradient of the stress tensor σ_{ij} as

$$\sigma_{ij,j} + X_i = 0 \qquad (i,j=1,2,3) \qquad (3.1)$$

Taking Hooke's law for the material to be given in terms of the shear modulus μ and Poisson's ratio ν, we obtain

$$\sigma_{ij} = \frac{2\mu\nu}{1-2\nu}\,\delta_{ij}\varepsilon_{mm} + 2\mu\varepsilon_{ij} \qquad (i,j,m=1,2,3) \qquad (3.2)$$

$$\varepsilon_{ij} = (u_{i,j} + u_{j,i})/2 \tag{3.3}$$

Combining eqs. (3.1)-(3.3), we obtain the Navier equations of equilibrium in terms of the displacements

$$u_{i,ij}/(1-2\nu) + \nabla^2 u_j + X_j/\mu = 0 \tag{3.4}$$

where ∇^2 is the usual Laplacian operator. Equation (3.4) is the governing set of partial differential equations to be satisfied in the body, denoted herein as **R**. In order to facilitate later solutions of eq. (3.4), we first seek to find a function of the displacements, such that the new governing equation lends itself more readily to analytical solution. The Galerkin vector G_i is one such function, although certainly not the only (see Pearson, 1959); the displacements are given in terms of derivatives of G_i as

$$u_i = \nabla^2 G_i - \frac{1}{2(1-\nu)} G_{j,ij} \tag{3.5}$$

The combination of terms in eq. (3.5) was originally selected such that substitution into the Navier equations (3.4) yields the following biharmonic equation

$$\nabla^2(\nabla^2 G_i) + X_i/\mu = 0 \tag{3.6}$$

Equation (3.6) applies for both the three-dimensional and two-dimensional problems.

Solution of eq. (3.6) for a specified body force provides, by the previous equations in this section, a solution of the stress state in **R**, subject to the imposed boundary conditions. In the general fracture mechanics problem, the necessary boundary conditions will be imposed on the regular surface **S** and on the crack surface Γ. The next section considers the solution to eq. (3.6) for the special or fundamental problems of point loads in infinite regions R^∞.

3.3 Fundamental Solutions

Fundamental solutions are singular solutions to eq. (3.6) for the infinite region. There are, in fact, an infinity of singular solutions of increasing order. Several of the lowest order singularities are the pole (point force), dipole (point twist), and the point pinch, depending on the application.

The properties of the singular solution play a crucial role in the BIE formulation. However, practical experience in terms of numerical solutions limits us to the use of the pole or point-force fundamental solution.

The point body force distribution will be taken to be represented by the Dirac delta function, as the amplitude times unit basis vectors (e_i=1,2,3), as follows

$$X_i = \delta(p-q)e_i \tag{3.7a}$$

Figure 3.1 illustrates the terminology to be used herein. The point force is applied at an internal point $p(\underset{\sim}{x})$; the response point is given as $q(\underset{\sim}{x})$.

Two of the necessary conditions for the representation given by eq. (3.7a) to be a point load at $p(\underset{\sim}{x})$ are

$$\begin{aligned} &1. \quad X_i \equiv 0 \qquad p \neq q \\ &2. \quad \int_{R^\infty} X_i \, dV = e_i \end{aligned} \tag{3.7b}$$

In the first of these, the body force distribution is identically zero, except at the source point for the load, $p(\underset{\sim}{x})$. Second, the net magnitude of the distribution is unity. Higher order terms such as net moments being zero are satisfied by the usual properties of the Dirac delta function.

The fundamental solution for the isotropic elastic problem is obtained by finding the solution to eq. (3.6) using eq. (3.7) for the body force. Denoting the fundamental Galerkin vector solution as G_i^* we obtain as the governing equation

$$\nabla^2(\nabla^2 G_i^*) + \delta(p-q)e_i/\mu = 0 \tag{3.8}$$

As discussed by Cruse (1977) and numerous earlier classical texts, the three-dimensional solution to eq. (3.8) is given by

$$G_i^* = \frac{1}{8\pi\mu} r(p,q) e_i \qquad (i=1,2,3) \qquad (3.9)$$

where r(p,q) is the scalar distance between $p(\underset{\sim}{x})$ and $q(\underset{\sim}{x})$, as shown in Fig. 3.1. The two-dimensional, fundamental Galerkin vector is given by

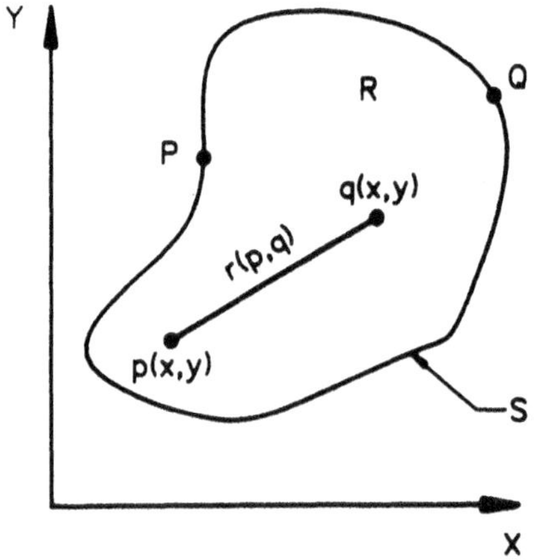

Figure 3.1. Some geometric definitions

$$G_i^* = \frac{1}{8\pi\mu} r^2 (1 + \log \frac{1}{r}) e_i \qquad (i=1,2) \qquad (3.10)$$

In the two-dimensional case, the r^2 term leads to a rigid body displacement; only the log(1/r) term actually gives rise to non-zero, singular stresses.

The displacements at the field point $q(\underset{\sim}{x})$ for the point load applied at $p(\underset{\sim}{x})$ are obtained by differentiation of the Galerkin vector representations of eqs. (3.9) and (3.10) according to eq. (3.5). Of course, the differentiation is at $q(\underset{\sim}{x})$, which is not taken to be equal to $p(\underset{\sim}{x})$. The derivatives of r(p,q) are then given by

$$r_{,i} = \partial r/\partial x_{i|q} = (x_{i|q} - x_{i|p})/r \tag{3.11}$$

$$= - \partial r/\partial x_{i|p}$$

The displacement solutions are found for three dimensions to be

$$u_i^*(p,q) = [(3-4\nu)\delta_{ij} + r_{,i}r_{,j}]e_j/[16\pi\mu(1-\nu)r(p,q)] \quad (i,j=1,2,3) \tag{3.12}$$

where δ_{ij} is the usual Kronecker delta; and, for two dimensions

$$u_i^*(p,q) = [(3-4\nu)\delta_{ij}\log(1/r) + r_{,i}r_{,j}]e_j/[8\pi\mu(1-\nu)] \quad (i,j=1,2) \tag{3.13}$$

These fundamental displacement solutions may be redefined in terms of the second-order tensor $U_{ji}(p,q)$ as

$$u_i^* (p,q) = U_{ji}(p,q)e_j \tag{3.14}$$

The displacements can be differentiated according to eq. (3.3) to obtain strains; these in turn are transformed through Hooke's law (3.2) into stress. The tractions along an arbitrary surface **S** within **R** with outward normal n_j, are given by

$$t_i = \sigma_{ij}n_j \tag{3.15}$$

such that the fundamental traction vector for three dimensions is given by

$$t_i^* = T_{ji}(p,q)e_j = - \{(dr/dn)[(1-2\nu)\delta_{ij} + 3r_{,i}r_{,j}] \tag{3.16}$$

$$- (1-2\nu)[n_jr_{,i} - n_ir_{,j}]\}e_j/[8\pi(1-\nu)r^2(p,q)]$$

and in two dimensions by

$$t_i^* = T_{ji}(p,q)e_j = -\{(dr/dn)[(1-2\nu)\delta_{ij} + 2r_{,i}r_{,j}] \tag{3.17}$$

$$- (1-2\nu)[n_j r_{,i} - n_i r_{,j}]\}e_j/[4\pi(1-\nu)r(p,q)]$$

It should be noted that all of the fundamental solutions are well behaved as $\nu \to 0.5$. In fact, experience shows that the numerical solution of the BIE (yet to be derived herein) is quite acceptable at $\nu = 0.5$.

Finally, it is easily shown that the fundamental traction solutions, eqs. (3.16) and (3.17), may be integrated over small spherical or circular surfaces S_ε near the point load point $p(\underset{\sim}{x})$. The result obtained is given for a small radius ε by

$$\lim_{\varepsilon \to o} \int_{S_\varepsilon} t_i^* \, dS = \delta_{ij} e_j \tag{3.18}$$

Thus, the net unit applied load solution is recovered.

3.4 Two-Dimensional Anisotropic Fundamental Solution

Consider first a rectilinear, anisotropic, elastic material with no symmetry in any of the three cartesian directions. It is well known (see Leknitskii, 1963) that there are 21 independent material elastic constants

$$\sigma_{ij} = C_{ijk\ell}\, \varepsilon_{k\ell} \qquad (i,\cdots,\ell = 1,2,3) \tag{3.19}$$

because

$$C_{ijk\ell} = C_{jik\ell} = C_{ij\ell k} = C_{k\ell ij} \tag{3.20}$$

It is convenient and usual to introduce a more compact notation, using symmetry of the stress and strain tensors. The independent components are arranged as a vector, such as shown for the stresses

$$\{\sigma_i\} = \begin{bmatrix} \sigma_1 & & \sigma_{11} \\ \sigma_2 & & \sigma_{22} \\ \sigma_3 & = & \sigma_{33} \\ \sigma_4 & & \sigma_{23} \\ \sigma_5 & & \sigma_{31} \\ \sigma_6 & & \sigma_{12} \end{bmatrix} \tag{3.21}$$

The Hooke's law relationship in terms of stiffnesses then becomes

$$\sigma_i = \hat{C}_{ij}\,\varepsilon_j \qquad (i,j=1,\cdots,6) \tag{3.22}$$

and in terms of compliances as

$$\varepsilon_i = \hat{S}_{ij}\,\sigma_j \qquad (i,j=1,\cdots,6) \tag{3.23}$$

Assume that the material is symmetric with respect to the x_3 direction. It is possible to show by reflecting the relations eq. (3.22) about x_3=0 for simple tensile stresses that

$$\hat{C}_{4i} \equiv \hat{C}_{5i} \equiv 0 \qquad (i=1,3) \tag{3.24}$$

and for simple shear stresses that

$$\hat{C}_{i4} \equiv \hat{C}_{i5} \equiv 0 \qquad (i=1,3) \tag{3.25}$$

For the case of an applied shear stress σ_{23} reflected about x_3=0, it can be shown that

$$\hat{C}_{46} \equiv \hat{C}_{56} \equiv 0 \tag{3.26}$$

Finally, the Hooke's law relationship for this state of symmetry has been reduced to thirteen constants

$$[\hat{C}_{ij}] = \begin{bmatrix} \hat{C}_{11} & \hat{C}_{12} & \hat{C}_{13} & 0 & 0 & \hat{C}_{16} \\ & \hat{C}_{22} & \hat{C}_{23} & 0 & 0 & \hat{C}_{26} \\ & & \hat{C}_{33} & 0 & 0 & \hat{C}_{36} \\ & \text{Symmetry} & & \hat{C}_{44} & \hat{C}_{45} & 0 \\ & & & & \hat{C}_{55} & 0 \\ & & & & & \hat{C}_{66} \end{bmatrix} \tag{3.27}$$

For plane strain or plane stress conditions, eq. (3.27) reduces further to needing six independent material constants. In the case of plane stress, the compliance relationship (3.23) becomes

$$\begin{Bmatrix} \varepsilon_x \\ \varepsilon_y \\ \gamma_{xy} \end{Bmatrix} = \begin{bmatrix} 1/E_{11} & -\nu_{12}/E_{11} & -\nu_{16}/E_{11} \\ -\nu_{21}/E_{22} & 1/E_{22} & -\nu_{26}/E_{22} \\ -\nu_{61}/G_{12} & -\nu_{62}/G_{12} & 1/G_{12} \end{bmatrix} \begin{Bmatrix} \sigma_x \\ \sigma_y \\ \tau_{xy} \end{Bmatrix} \tag{3.28}$$

For the more usual case of an orthotropic layer, $\nu_{16}=\nu_{26}=0$ and $\nu_{61}=\nu_{62}=0$. The remainder of the constants in eq. (3.28) are the usual engineering constants of Young's modulus and Poisson's ratio.

The fundamental solution for the two-dimensional problem with material symmetry about $x_3=0$ is easily obtained from an Airy stress function formulation of the problem. Following the usual approach (Leknitskii, 1963), equilibrium is satisfied by taking stresses in terms of derivatives of the stress function F(x,y) as

$$\sigma_x = \partial^2 F/\partial y^2 \; ; \; \sigma_y = \partial^2 F/\partial x^2 \; ; \; \tau_{xy} = -\partial^2 F/\partial x \partial y \tag{3.29}$$

The equation of compatibility in terms of the strains is given as

$$\frac{\partial^2 \varepsilon_x}{\partial y^2} + \frac{\partial^2 \varepsilon_y}{\partial x^2} = \frac{\partial^2 \gamma_{xy}}{\partial x \partial y} \tag{3.30}$$

Hooke's law for the compliance form (3.28) will be taken in the following manner

$$\varepsilon_i = \beta_{ij}\sigma_j \qquad (i,j=1,2,6) \tag{3.31}$$

Combining these last three equations, we obtain the governing equation for F(x,y) as

$$\beta_{22}\frac{\partial^4 F}{\partial x^4} - 2\beta_{26}\frac{\partial^4 F}{\partial x^3 \partial y} + (2\beta_{12} + \beta_{66})\frac{\partial^4 F}{\partial x^2 \partial y^2}$$
$$- 2\beta_{16}\frac{\partial^4 F}{\partial x \partial y^3} + \beta_{11}\frac{\partial^4 F}{\partial y^4} = 0 \tag{3.32}$$

We now seek a solution to (3.22) in terms of a complex coordinate

$$z = x + \mu y \tag{3.33}$$

where

$$\mu = a + ib \qquad ; \; i=\sqrt{-1} \tag{3.34}$$

Substituting eq. (3.34) into (3.32) we obtain the characteristic equation for μ as

$$[\beta_{22} - 2\mu\beta_{26} + (2\beta_{12}+\beta_{66})\,\mu^2 - 2\beta_{16}\mu^3 + \beta_{11}\mu^4]\,\frac{d^4F}{dz^4} = 0 \tag{3.35}$$

For nontrivial solutions to the stress function formulation, the term in square brackets must be identically zero. The zeros of this characteristic equation are distinct, so long as the material is anisotropic. It may be shown that to have a positive definite strain energy the roots of eq. (3.35) must be imaginary or complex (conjugate pairs). As an example, if we take the orthotropic plane stress problem (see eq. 3.28) and assign engineering constants as follows,

$$\begin{aligned} E_{11} &= 30 \times 10^6 \text{ psi} \\ E_{22} &= 3 \times 10^6 \text{ psi} \\ G_{12} &= 1 \times 10^6 \text{ psi} \\ \nu_{12} &= 0.30 \end{aligned} \tag{3.36}$$

then the following roots to eq. (3.35) are obtained

$$\mu_1 = -\mu_2 = 0.573i \qquad (3.37)$$
$$\mu_3 = -\mu_4 = 5.44i$$

where again $i = \sqrt{-1}$.

The most general form of the stress function F(x,y) for the anisotropic problem may be expressed in terms of two complex coordinates

$$z_k = x + \mu_k y \qquad (k=1,2) \qquad (3.38)$$

where $\mu_k, \bar{\mu}_k$ are the four complex roots of eq. (3.35). The bar on the second root denotes the conjugate root. The stress function may then be written as (Re denotes the real part)

$$F(z_1,z_2) = F_1(z_1) + F_2(z_2) + \overline{F_1(z_1)} + \overline{F_2(z_2)} \qquad (3.39)$$
$$= 2\mathrm{Re}[F_1(z_1) + F_2(z_2)]$$

The stresses corresponding to this stress function are then given by differentiation of eq. (3.39) according to the definitions in eq. (3.29) such that

$$\sigma_x = 2\mathrm{Re}\left[\mu_1^2 \frac{d^2F_1}{dz_1^2} + \mu_2^2 \frac{d^2F_2}{dz_2^2}\right]$$

$$\sigma_y = 2\mathrm{Re}\left[\frac{d^2F_1}{dz_1^2} + \frac{d^2F_2}{dz_2^2}\right] \qquad (3.40)$$

$$\tau_{xy} = -2\mathrm{Re}\left[\mu_1 \frac{d^2F_1}{dz_1^2} + \mu_2 \frac{d^2F_2}{dz_2^2}\right]$$

The displacements corresponding to these stresses can be obtained by application of Hooke's law followed by one integration of the stress functions. If we let $\Phi_i = dF_i/dz_i$ (no summation on the subscript -i), then it is possible to show that, within rigid body terms, the x- and y-displacements are

$$u(x,y) = 2\mathrm{Re}[p_1\Phi_1 + p_2\Phi_2] \tag{3.41}$$

$$v(x,y) = 2\mathrm{Re}[q_1\Phi_1 + q_2\Phi_2]$$

where

$$p_k = \beta_{11}\mu_k^2 + \beta_{12} - \beta_{16}\mu_k$$

and

$$q_k = \beta_{22}/\mu_k + \beta_{12}\mu_k - \beta_{26}. \tag{3.42}$$

If we consider a closed contour S surrounding the point load for the fundamental solution, the net force on S is given by the following components, P_x and P_y

$$P_x = \sum_{k=1}^{2} [[\mu_k\Phi_k + \bar{\mu}_k\bar{\Phi}_k]]$$

$$P_y = - \sum_{k=1}^{2} [[\Phi_k + \bar{\Phi}_K]] \tag{3.43}$$

where the double brackets denote the jump in the function for a closed contour surrounding the point load.

To ensure that the fundamental solution has single valued displacements, it is necessary that

$$[[\mathrm{Re}\ (\sum_{k=1}^{2} p_k\Phi_k)]] \equiv 0$$

$$[[\mathrm{Re}\ (\sum_{k=1}^{2} q_k\Phi_k)]] \equiv 0 \tag{3.44}$$

The fundamental solutions for the anisotropic two-dimensional problems are given in terms of the following logarithmic functions

$$\Phi_{ik} = A_{ik}\log(z_k) \tag{3.45}$$

where the index i refers to the component of the applied load. If we take P_x=P_y=1, then eq. (3.43) results in the following

$$\sum_{k=1}^{2} (A_{jk} - \bar{A}_{jk}) = \delta_{j2}/2\pi i$$
$$- \sum_{k=1}^{2} (\mu_k A_{jk} - \bar{\mu}_k \bar{A}_{jk}) = \delta_{j1}/2\pi i \qquad (3.46)$$

Continuity of the displacements for the fundamental solutions is satisfied by the following relations

$$\sum_{k=1}^{2} (p_k A_{jk} - \bar{p}_k \bar{A}_{jk}) = 0$$
$$\sum_{k=1}^{2} (q_k A_{jk} - \bar{q}_k \bar{A}_{jk}) = 0 \qquad (3.47)$$

where p_k, q_k are given by eq. (3.42).

The equations, (3.46) and (3.47), constitute a sufficient set of equations for the complex constants A_{jk}. In turn, these constants define the fundamental solution, eq. (3.45), for which the stresses are given by eq. (3.40) and the displacements by eq. (3.41).

The reduction of the anisotropic fundamental solution to that for the isotropic case can be made. In order to do so it is necessary to expand the characteristic equation, (3.32), in terms of a perturbed isotropic material. A satisfactory perturbation is to take the shear modulus to be nearly equal to its isotropic value.

$$G_{12} = (1-\varepsilon)\, E/[2(1+\nu)] \qquad (3.48)$$

where ε is a small number; and to take $E_{11} = E_{22} = E$, and $\nu_{12} = \nu$.

3.5 Three-Dimensional Anisotropic Fundamental Solution

The three-dimensional fundamental solution for the generally anisotropic material cannot be developed in closed form. The basic theoretical development follows that of F. John (1955) and of Vogel and Rizzo (1973). The actual algorithm to be described was developed by Wilson and Cruse (1978).

Based on the work of F. John, the fundamental displacement solution may be written as

$$U_{ij}(p,q) = \frac{1}{8\pi^2 r(p,q)} \oint_{|\xi|=1} K_{ij}^{-1}(\underset{\sim}{x})\, ds \tag{3.49}$$

where r(p,q) is as previously defined, and $K_{ij}(\underset{\sim}{x})$ is given in the global coordinates x_i as

$$K_{ij}(\underset{\sim}{x}) = C_{ijkm} x_k x_m \tag{3.50a}$$

The quadratic polynomial in eq. (3.50) is formed from the stiffness matrix of eq. (3.19). The path integral in eq. (3.49) is integrated on a unit circle in a plane normal to the line between p,q.

In the case of an isotropic material, eq. (3.50) may be written as

$$K_{ij}(\underset{\sim}{x}) = 2\mu\left[\frac{\nu}{1-2\nu}\,\delta_{ij} x_m x_m + x_i x_j\right] \tag{3.50b}$$

The polynomial for the isotropic material can be inverted to form

$$K_{ij}^{-1}(\underset{\sim}{x}) = \left[\delta_{ij} - x_i x_j/2(1-\nu)\right]/\mu \tag{3.51}$$

Substitution of eq. (3.51) into the path integral in eq. (3.49) results in the original three-dimensional displacement fundamental solution, given by eq. (3.12).

In the case of anisotropic materials, the integral in eq. (3.49) must be approximated.* The first step is to transform the modulation kernel,

---

* The only exception is transversely isotropic material, as shown by Pan and Chou (1976).

eq. (3.49), into a local coordinate system. The new coordinate system **y** is constructed such that the y_3 axis is along the p,q line; y_1,y_2 are in the plane of the unit circle. The transformation may be written as

$$y_i = a_{ij}x_j \tag{3.52}$$

in the usual sense of cartesian transformations. The parameters of a_{ij} may conveniently be taken as Φ,Θ, the spherical coordinates of the line segment.

The fundamental traction solution and other higher derivatives are also needed. The first derivative of the modulation kernel in eq. (3.49) is given by

$$K^{-1}_{ij,k} = K^{-1}_{ij,\alpha}\nu_{\alpha,k} \tag{3.53}$$

where $\nu_\alpha=(\Phi,\Theta)$. Further, for convenience, the inverse form in eq. (3.53) can be calculated from the following identity

$$K_{ij}\,K^{-1}_{jk} = \delta_{ik} \tag{3.54}$$

such that

$$K^{-1}_{ij,\alpha} = -\,K^{-1}_{ik}\,K_{k\ell,\alpha}K^{-1}_{\ell j} \qquad (\alpha=1,2) \tag{3.55}$$

Equation (3.55) makes use of analytical differentiation of the kernel $K_{k\ell}(\nu_\alpha)$,obtained by substituting the spherical coordinate transformation (3.52) into the quadratic form (3.50).

The fundamental solution and its derivatives are obtained by a numerical integration of eq. (3.47) and its derivatives around the unit circle. Since the integrand depends on the orientation of the p,q line, it is convenient to calculate these results for a given anisotropic material in a tabular form, for the full range of values of ν_α. Wilson and Cruse (1978) detail this procedure and show that good accuracy is achieved with relatively little computational effort.

The table of values for the modulation integral in eq. (3.49) and for its derivatives forms the basis for evaluation the fundamental solutions. As the line p,q changes orientation during the actual computation of the BIE model, the values of the modulation integral are obtained by interpolation on ν_α. Again, as shown by Wilson and Cruse (1978), this approach leads to fundamental solutions with satisfactory accuracy for computational modeling.

3.6 Somigliana Identities

The reciprocal work theorem of Betti (1872) forms the basis of the BIE formulation. The reciprocal work theorem begins with the following identity for the elastic problem

$$\int_V \sigma_{ij}^1 \, \varepsilon_{ij}^2 \, dV = \int_V \sigma_{ij}^2 \, \varepsilon_{ij}^1 \, dV \tag{3.56}$$

where the superscripts denote two different solution states. The validity of the identify is easily established by substituting the form of Hooke's law from eq. (3.19) for stress state-1

$$\int_V \sigma_{ij}^1 \varepsilon_{ij}^2 \, dV = \int_V C_{ijk\ell} \varepsilon_{k\ell}^1 \, \varepsilon_{ij}^2 \, dV \tag{3.57}$$

Reversal of the strain terms is achieved by recognizing the symmetry $C_{ijk\ell} = C_{k\ell ij}$ such that eq. (3.57) becomes

$$\int_V \sigma_{ij}^1 \, \varepsilon_{ij}^2 \, dV = \int_V \sigma_{k\ell}^2 \varepsilon_{k\ell}^1 \, dV \quad , \tag{3.58}$$

completing the proof.

Substitution of the displacement gradient tensors for the strain tensors in eq. (3.58) then gives

$$\int_V \sigma_{ij}^1 u_{i,j}^2 \, dV = \int_V \sigma_{ij}^2 \, u_{i,j}^1 \, dV \tag{3.59}$$

Let state-1 be a well-posed boundary value problem and let state-2 be the fundamental solutions. Integrating first by parts we obtain from eq. (3.59)

$$\int_V \sigma_{ij} \, u_{i,j}^* dV = \int_V (\sigma_{ij} \, u_i^*)_{,j} \, dV - \int_V \sigma_{ij,j} u_i^* \, dV \tag{3.60}$$

and then apply the divergence theorem to obtain

$$\int_V \sigma_{ij} u^*_{i,j} dV = \int_S t_i u^*_i \, dS + \int_V X_i u_i^* \, dV \tag{3.61}$$

The usual formulation will be for $X_i \equiv 0$, so that the last term in eq. (3.61) may be dropped. Applying the same integration by parts with the divergence theorem on the right-hand side of eq. (3.59), and recalling the singular property of the fundamental solution in eq. (3.7), we obtain

$$\int_V \sigma^*_{ij} u_{i,j} \, dV = \int_S t^*_i u_i \, dS + \int_V \delta(p-q) \, e_i \, u_i \, dV \tag{3.62}$$

The properties of the Dirac body force term in eq. (3.62) reduce the volume integral to a free-term expression involving u_i evaluated at the source point $p(\underset{\sim}{x})$

$$\int_V \sigma_{ij}{}^* \, u_{i,j} \, dV = \int_S t_i{}^* \, u_i \, dS + u_i e_i \tag{3.63}$$

Using the terminology for the fundamental solutions given in eqs. (3.14) and (3.16), and slightly rearranging terms, the Somigliana (1885) identity is obtained

$$u_i(p) = \int_S U_{ij}(p,Q) t_j(Q) \, dS - \int_S T_{ij}(p,Q) u_j(Q) \, dS \tag{3.64}$$

The boundary points in eq. (3.64) and in what follows are denoted by upper case letters.

The function of the fundamental solutions in the formulation of integral operator solutions to elasticity problems may now be elucidated. Equation (3.64) states that the interior displacements at an arbitrary point are now known to depend on surface data only. Further, eq. (3.64) is differentiable; that is, since $p(\underset{\sim}{x})$ is an arbitrary point inside the body, derivatives, of eq. (3.64) may be computed as follows

$$u_{i,k}(p) = \int_S \frac{\partial U_{ij}(p,Q)}{\partial x_{k|p}} t_j(Q) \, dS - \int_S \frac{\partial T_{ij}(p,Q)}{\partial x_{k|p}} u_j(Q) \, dS \tag{3.65}$$

Since the kernels $U_{ij}(p,Q)$ and $T_{ij}(p,Q)$ in eq. (3.65) depend only on the distance $r(p,Q)$ and its derivatives and given that

$$\frac{\partial r}{\partial x_{k|p}} = - \frac{\partial r}{\partial x_{k|q}} \tag{3.66}$$

we can write eq. (3.65) as follows

$$u_{i,k}(p) = -\int_S U_{ij,k}(p,Q)t_j(Q)\,dS + \int_S T_{ij,k}(p,Q)u_j(Q)\,dS \tag{3.67}$$

The derivatives of the kernels in eq. (3.67) and in subsequent formulas will be assumed to be with respect to $Q(\underset{\sim}{x})$, except when otherwise given.

Application of Hooke's law to eq. (3.67) results in the Somigliana identity for the interior stresses

$$\sigma_{ik}(p) = \int_S U^{\sigma}_{ikj}(p,Q)t_j(Q)\,dS - \int_S T^{\sigma}_{ikj}(p,Q)u_j(Q)\,dS\,. \tag{3.68}$$

The detailed forms of the kernels for the various isotropic cases are given by Rizzo (1967) and Cruse (1969), for the two-dimensional anisotropic formulation by Snyder and Cruse (1975), and implicitly, for three-dimensional anisotropic elasticity by Wilson and Cruse (1978).

The point of greatest importance to a proper understanding of the integral formulation and the role of the point-load fundamental solution is that the stresses in eq. (3.68) fully satisfy equilibrium at all interior points, irrespective of boundary modeling. This essential conclusion comes from the property of differentiability in eqs. (3.64) and (3.68), and the fact that the fundamental solutions, or kernels U_{ij} and T_{ij}, are each equilibrium solutions. Application of the equilibrium operation in eq. (3.1) to eq. (3.68) results in the following identities

$$\begin{aligned} U^{\sigma}_{ikj,k}(p,Q) &\equiv 0 \qquad p\neq Q \\ T^{\sigma}_{ikj,k}(p,Q) &\equiv 0 \qquad p\neq Q \end{aligned} \tag{3.69}$$

Thus, the fundamental solution using the point-load representation leads to the Somigliana identities, eqs. (3.64) and (3.68), with the key solution variables on the interior as free variables. Then, the properties of the fundamental solution kernels in these integral identities guarantee satisfaction of internal equilibrium.

3.7 Boundary-Integral Equations

The Somigliana identities do not, of course, represent a solution to a boundary value problem. These identities require for evaluation that the totality of boundary data be specified--the tractions and displacements at all surface points. The critical and significant contributions of Jaswon and Ponter (1963) and Rizzo (1967) was the simple operation of taking the interior point to the surface in eq. (3.64), or its equivalent in potential theory.

In essentially all of the integral equation formulations for potential theory and elasticity to that time and ongoing today, these early developments all led into the cul-de-sac of the Fredholm theorems. The historical break of Jaswon and Rizzo cannot be overstated, as the classical efforts resulted in formulations which could not be applied practically to physical problems involving bodies which had non-smooth surfaces.

A proper discussion of the two paths of integral equation developments cannot be given here. In order to properly distinguish between the two developments, Cruse (1969) denoted the classical efforts as <u>indirect</u> potential methods, and those leading to the BIE as <u>direct</u> potential methods. The notion of direct has to do with the nature of the unknowns in the BIE and will be discussed further below. A more detailed derivation and discussion of both approaches is given in Cruse (1977).

The now well-known discontinuity property in the Somigliana identity (3.64) is now briefly reviewed. Let the surface of the body be separated into two surfaces--one a small region S_ε surrounding a surface point denoted $P(\underset{\sim}{x})$; the second surface is then $S-S_\varepsilon$. Equation (3.64) may be written then as follows

$$
\begin{aligned}
u_i(p) = \int_{S-S_\varepsilon} U_{ij}(p,Q)t_j(Q)dS - \int_{S-S_\varepsilon} T_{ij}(p,Q)\, u_j(Q)\, dS \\
+ \int_{S_\varepsilon} U_{ij}(p,Q)t_j(Q)\, dS - \int_{S_\varepsilon} T_{ij}(p,Q)u_j(Q)\, dS
\end{aligned}
\tag{3.70}
$$

All of the integrals on $S-S_\varepsilon$ are regular as $p(\underset{\sim}{x}) \to P(\underset{\sim}{x})$, and the size parameter ε is taken in the limit to zero. These integrals are then known as principal value integrals. The principal value integral on S is denoted as ⟨S⟩.

Consider now the integrals on S_ε. The order of the singularity due to $r(p,P) \to 0$ in the fundamental solution kernels in eq. (3.70) for three-dimensional problems is $1/\varepsilon$ and $1/\varepsilon^2$, respectively. Since the surface area of the deleted region is necessarily of order ε^2, the integral of $U_{ij}(p,P)$ in the limit on S_ε as $\varepsilon \to 0$ is zero.

To obtain the limiting form for the last integral in eq. (3.70), the integral can be regularized as follows

$$\int_{S_\varepsilon} T_{ij}(p,Q)u_j(Q)dS = \int_{S_\varepsilon} T_{ij}(p,Q)[u_j(Q)-u_j(P)]\,dS + u_j(P)\int_{S_\varepsilon} T_{ij}(p,Q)\,dS \tag{3.71}$$

The effect of regularization is that the term $[u_i(Q)-u_i(P)]$ is of order ε due to continuity of the displacements; thus, the integral of the bracketed term in eq. (3.71) is of order ε and vanishes in the limit as $\varepsilon \to 0$.

Thus, only the last term in eq. (3.71) is of importance. Taking the distance from $p(\tilde{x})$ to $P(\tilde{x})$ to be δ, where $\delta \ll \varepsilon$, the result can in many cases be directly computed. For smooth surfaces, Cruse (1977) details the integrations in two and three dimensions; Cruse et al. (1977) detail the axisymmetric results. In general, we may replace the last integral in eq. (3.71) by its limiting value

$$C_{ij}(P) = \lim_{\varepsilon \to 0} \{\lim_{\delta \to 0} \int_{S_\varepsilon} T_{ij}(p,Q)\,dS\} \tag{3.72}$$

The well-known result for flat surfaces is that $C_{ij}(P) = \delta_{ij}/2$.

When the surface is not flat, the value of $C_{ij}(P)$ may be found analytically in two dimensions; the three-dimensional result is not easily obtained in a general way. Cruse (1974) introduced the simple device of equilibrium for the fundamental solution to write

$$C_{ij}(P) = -\int_{\langle S \rangle} T_{ij}(P,Q)\,dS \tag{3.73}$$

In general, the principal value integral in eq. (3.73) may be computed with good accuracy using data which is readily available in the numerical quadrature of the BIE.

The boundary-integral equation for linear elasticity may now be written

$$C_{ij}u_j(P) + \int_{\langle S\rangle} T_{ij}(P,Q)\, u_j(Q)\, dS = \int_{\langle S\rangle} U_{ij}(P,Q)t_j(Q)\, dS \qquad (3.74)$$

The BIE is a formal constraint equation between the physical boundary data. The BIE is a direct means for computing unknown boundary data from known boundary data. As briefly stated in the previous section, the direct solution method for integral equation models offers overwhelming advantages over the classical, indirect methods for the usual engineering geometries with corners and cracks.

3.8 Numerical Quadrature of the BIE

Reduction of the BIE to an algebraic form requires that the boundary data be approximated in some systematic manner. The approximations first involve replacing the boundary by a complete set of surface patches or segments (these are conveniently called boundary elements). The boundary data is then interpolated over each boundary element, thereby reducing eq. (3.74) to algebraic form.

Two approaches will be used in this monograph, reflecting the two principal computer codes that will be illustrated. Each of the two has its own advantages and limitations. In the first, the boundary elements are taken to be flat (planes or lines) and the boundary data is linearly interpolated over the segments. This approach will be illustrated for the two-dimensional problem.

Figure 3.2 shows one-quarter of a plate with a hole. Linear boundary elements are shown for the modeled surfaces. The horizontal x-axis is a plane of symmetry, as is the vertical y-axis. Two symmetry modeling approaches will be described. The surfaces to be modeled are represented by the nodes shown, connected by straight-line segments. The hole is modeled with emphasis near the horizontal axis where the stress concentration is of interest.

The vertical axis of symmetry is treated as a modeled surface with roller boundary conditions used to satisfy the symmetry conditions for zero normal displacements and zero shear stresses. The horizontal symmetry line is not modeled; rather, the entire model for $y>0$ is reflected about the x-axis by the computer algorithm in order to generate the entire modeled surface. However, the number of unknowns is set by the nodes shown in Fig. 3.2, as the solution is also symmetric. This symmetry algorithm is usually available in all modern BEM codes.

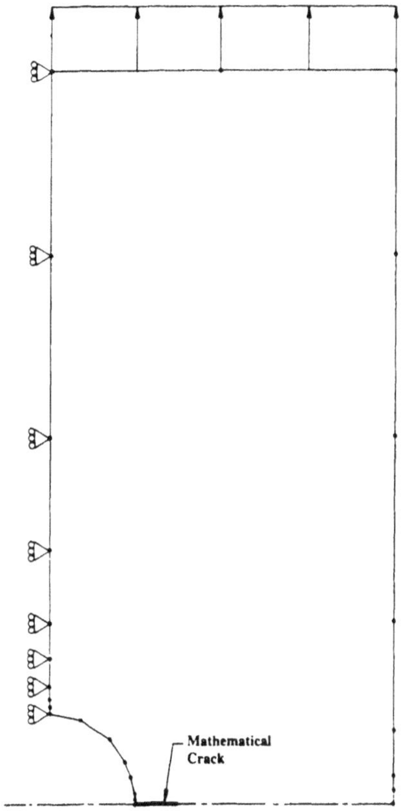

Figure 3.2. BEM mesh for piecewise linear model of notched plate

Figure 3.3 illustrates the view of one boundary segment containing the integration point $Q(\underset{\sim}{x})$, as seen from $P(\underset{\sim}{x})$. The approach in building an algebraic system to model eq. (3.74) will be to select $P(\underset{\sim}{x})$, and then to integrate the $Q(\underset{\sim}{x})$ over all of S; this is followed by moving to the next $P(\underset{\sim}{x})$ and repeating the integration process.

Let us interpolate the boundary data on the element shown as a linear interpolation. For convenience, let $\underset{\sim}{Q},\underset{\sim}{P}$ be vectors defining the point positions relative to a global coordinate system. The origin of that system may be moved to $P(\underset{\sim}{x})$ without any loss of generality. Finally, let any boundary data be given as the vector quantity

$$f_i(Q) = \sum_{m=1}^{2} \{[\frac{1}{2} - (-1)^m \frac{\bar{\underset{\sim}{Q}}}{\underset{\sim}{\Delta S}} + (-1)^m \frac{\underset{\sim}{Q}}{\underset{\sim}{\Delta S}}]f_i^{\,m}\} \qquad (3.75)$$

where the centroid $\bar{\underset{\sim}{Q}}$ and arc length vector $\Delta\underset{\sim}{S}$ are given by

$$\bar{\underset{\sim}{Q}} = \frac{1}{2}(\underset{\sim}{Q}_1+\underset{\sim}{Q}_2)$$

$$(3.76)$$

$$\Delta\underset{\sim}{S} = \underset{\sim}{Q}_2-\underset{\sim}{Q}_1$$

and where $f_i^{\,m}$ is the set of nodal values at the two ends of the line segment.

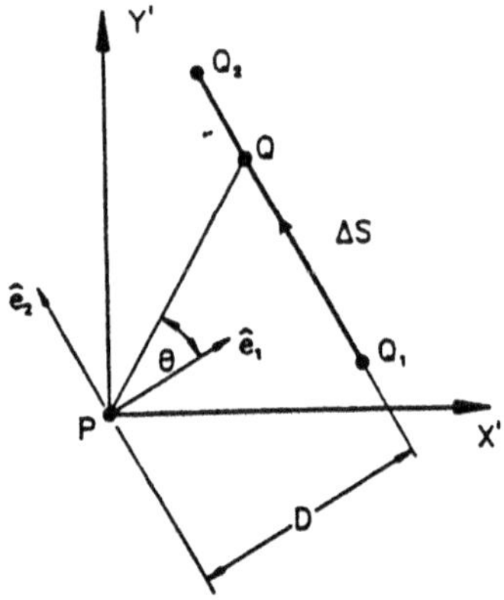

Figure 3.3. Single boundary segment

Equation (3.75) may be understood by moving $\underset{\sim}{Q}$ to the end segment points $\underset{\sim}{Q}_1$ and $\underset{\sim}{Q}_2$ and finding

$$f_i(Q_m) = f_i^{\,m} \qquad (3.77)$$

Further, the interpolation algorithm in eq. (3.75) contains two constant terms and a linear term. Each square bracketed term must be integrated against the kernels in eq. (3.74) to obtain an algebraic system

$$C_{ij}u_j(P_N) + \sum_{M=1}^{NSEG} \int_{\Delta S_M} \{ \sum_{m=1}^{2} ([\frac{1}{2} - (-1)^m \frac{\bar{\underset{\sim}{Q}}}{\underset{\sim}{\Delta S}} + (-1)^m \frac{\underset{\sim}{Q}}{\underset{\sim}{\Delta S}}]\ u_j^m)\ T_{ij}(P_N,Q)\ dS\}$$

$$= \sum_{M=1}^{NSEG} \int_{\Delta S_M} \{ \sum_{m=1}^{2} ([\frac{1}{2} - (-1)^m \frac{\bar{\underset{\sim}{Q}}}{\underset{\sim}{\Delta S}} + (-1)^m \frac{\underset{\sim}{Q}}{\underset{\sim}{\Delta S}}]\ t_j^m)\ U_{ij}(P_N,Q)\ dS\} \tag{3.78}$$

The three-dimensional algorithm equivalent to eq. (3.78) is given in Cruse (1974).

The terms in eq. (3.78) require integration of the kernel functions times constant and linear variation coordinate terms. For flat segments in two and three dimensions, these integrals have been obtained in closed form.

The traction components in eq. (3.78) must be allowed to be discontinuous at nodes in order to model sensible engineering problems. While some have dealt with this issue through the use of double nodes, a much simpler algorithm suffices.

Essentially, the approach is to treat displacements as nodal variables in the storage algorithm. This results in continuity of the boundary displacements. The boundary tractions on the other hand are treated as element variables, having distinct values at each end of the corner of the boundary element.

If the tractions are discontinuous, the discontinuous values are either specified, by reference to the appropriate segment, or at most one of the two values may be unknown. The input data logic requires only that information defining which segment carries the unknown traction be supplied. If the unknown traction is shared by two segments, the tractions are continuous. These conditions fully reflect physically plausible boundary conditions for tractions.

If we let the values of the kernel integrals in eq. (3.78) be denoted as ΔU_{ij} and ΔT_{ij} then the full algebraic system of equations becomes

$$\underset{(2N\times 2N)}{[C_{ij}]}\ \underset{(2N\times 1)}{\{u_j\}} + \underset{(2N\times 2N)}{[\Delta T_{ij}]}\ \underset{(2N\times 1)}{\{u_j\}} = \underset{(2N\times 4M)}{[\Delta U_{ij}]}\ \underset{(4M\times 1)}{\{t_j\}} \tag{3.79}$$

In the two-dimensional problem of eq. (3.79), N is the number of nodes and M is the number of segments, with two ends per segment. When there is symmetry, the number of columns is doubled for each degree of symmetry (1 or 2).

In actual practice, the full ΔT_{ij} and ΔU_{ij} matrices are never stored according to these sizes. As the system of equations is generated, the known boundary conditions are multiplied out to generate a right-hand-side (RHS) vector; the coefficients of the unknown terms {x} populate a coefficient matrix [A] such that

$$\underset{(2N\times 2N)}{[A_{ij}]}\ \underset{(2N\times 1)}{\{x_j\}} = \underset{(2N\times 1)}{\{RHS\}} \tag{3.80}$$

is the working form of eq. (3.79). Equation (3.80) is solved by normal matrix reduction methods.

The second approach, first used by Lachat and Watson (1976), is to model the surface geometry and the boundary data by quadratic isoparametric interpolation functions. In this case the boundary coordinates and boundary data are given on a given boundary element in terms of shape-functions $N_\alpha(\underset{\sim}{\xi})$

$$\begin{aligned} x_i(Q) &= N_\alpha x_i^\alpha \\ u_i(Q) &= N_\alpha u_i^\alpha \\ t_i(Q) &= N_\alpha t_i^\alpha \end{aligned} \tag{3.81}$$

In eq. (3.81), α ranges over the number of nodes for a given element. In two-dimensional elements with quadratic data the range is three; in three-dimensional surface elements with quadratic data the range is eight for quadrilateral elements and six for triangular elements. The shape functions in eq. (3.81) are given in eq. (4.20).

Application of the isoparametric modeling for the general boundary value problem follows the same logic as used above for linear variation.

This logic includes that for discontinuous tractions and for symmetry. The system equations (3.79) and (3.80) also hold.

Figure 3.4 illustrates in a three-dimensional stress analysis model of a compound notch (a), equivalent boundary element models using linear flat triangles (b), and quadratic isoparametric elements (c). Table 3.1 summarizes the computer run times and modeling data for the two cases. The linear code BINTEQ uses exact integration for the linear interpolations. The quadratic BIE code can also use linear interpolation by suppressing the mid-side node freedoms. Those results, while the least expensive, are also the least accurate. The extra effort required to generate the numerical integrals for the quadratic data model seems justified by the difference in the predicted stress concentrations.

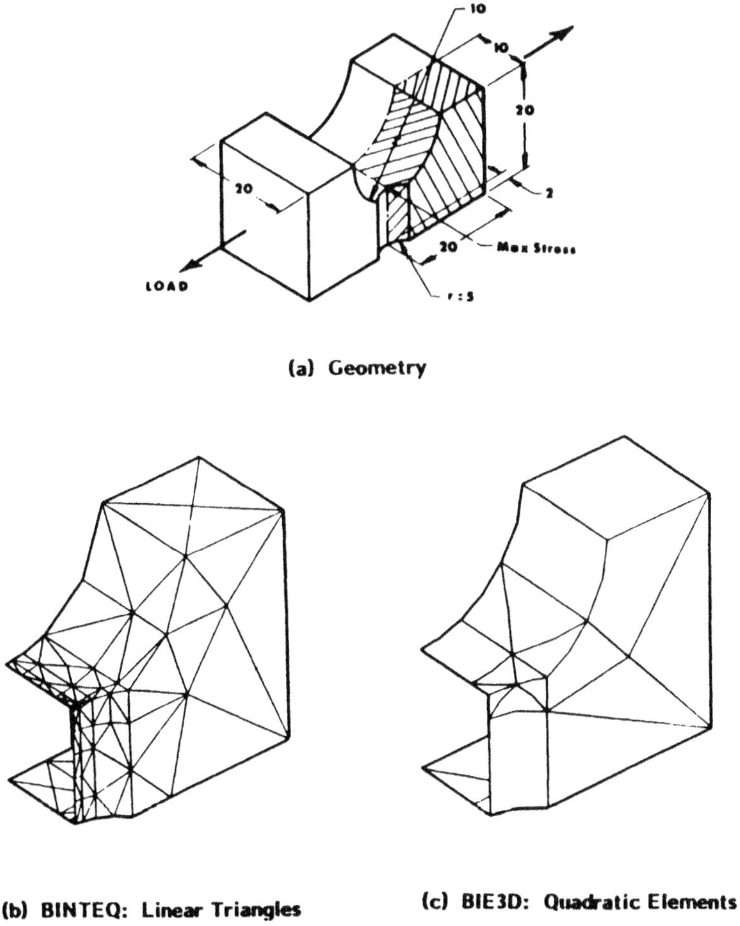

Figure 3.4. Structural detail with intersecting stress concentrations

Table 3.1. Three-dimensional notch model data

	BINTEQ (Linear)	(Linear)	(Quadratic)
No. of Elements	115	24	24
No. of Nodes	70	23	69
Planes of Symmetry	2	2	2
Degrees-of-Freedom	840	276	828
CPU Seconds (VAX)	1578	1045	3404
K_T	7.07*	6.71**	7.63**

* Based on the element centroidal value of stress.
** Based on the nodal value of stress.

References

(1872) E. Betti, Teoria dell Elasticita, Il Nuovo Ciemento, t.7-10.

(1885) C. Somigliana, Sopra l'equilibriodi un Corpo Elastico Istropo, Il Nuovo Ciemento, t.17-19.

(1953) O.D. Kellogg, Foundations of Potential Theory, Dover.

(1955) F. John, Plane Waves and Spherical Means Applied to Partial Differential Equations, Interscience.

(1959) C.E. Pearson, Theoretical Elasticity, Harvard University Press, Cambridge, Mass.

(1963) M.A. Jaswon and A.R. Ponter. An Integral Equation Solution of the Torsion Problem, Proceedings of the Royal Society, A, 237-246.

(1963) S.G. Lekhnitskii, Theory of Elasticity of an Anisotropic Elastic Body, Holden Day, San Francisco.

(1965) V.D. Kupradze, Potential Methods in the Theory of Elasticity, D. Davey.

(1967) F.J. Rizzo, An Integral Equation Approach to Boundary Value Problems of Classical Elastostatics, Quarterly of Applied Mathematics, 25, 83-96.

(1969) T.A. Cruse, Numerical Solutions in Three-Dimensional Elastostatics, International Journal of Solids and Structures, 5, 1259-1274.

(1973) S.M. Vogel and F.J. Rizzo, An integral equation formulation of three dimensional anisotropic elastostatic boundary value problems, Journal of Elasticity, 3, 3, 001-014.

(1974) T.A. Cruse, An Improved Boundary-Integral Equation Method for Three Dimensional Elastic Stress Analysis, Computers and Structures, 4, 741-754.

(1975) M.D. Snyder and T.A. Cruse, Boundary-integral Equation Analysis of Cracked Anisotropic Plates, A Contour Integral Computation of Mixed-Mode Stress Intensity Factors, International Journal of Fracture, 11, 315-328.

(1976) J.C. Lachat and J.O. Watson, Effective Numerical Treatment of Boundary Integral Equations: A Formulation for Three-Dimensional Elastostatics, International Journal for Numerical Methods in Engineering, 10, 991-1005.

(1976) Y.C. Pan and T.W. Chou, Point Force Solution for an Infinite Transversely Isotropic Solid, ASME Paper No. 76-WA/APM-18, New York.

(1977) T.A. Cruse, Mathematical Foundations of the Boundary-Integral Equation Method in Solid Mechanics, AFOSR-TR-77-1002 (Reprinted by Southwest Research Institute, San Antonio).

(1977) T.A. Cruse, D.W. Snow, and R.B. Wilson, Numerical Solutions in Axisymmetric Elasticity, Computers and Structures, 7, 445-451.

(1978) R.B. Wilson and T.A. Cruse, Efficient Implementation of Anisotropic Three Dimensional Boundary-Integral Equation Stress Analysis, International Journal for Numerical Methods in Engineering, 12, 1383-1397.

4.0 BIE Modeling of Crack Surfaces

4.1 Introduction

The boundary-integral equation formulation of elasticity problems was shown in Chapter 3 to have the powerful characteristic of being able to fully satisfy interior equilibrium, for a given boundary solution. The source of modeling error for BEM analyses is therefore not the volume discretization, as in the FEM, but rather the boundary discretization.

The essential closeness of the BIE formulation to the full equilibrium differential equations also contrasts with FEM formulations which rely on variational or Galerkin methods. This closeness results in a major limitation for BEM analysis of fracture mechanics problems: the BIE formulation degenerates for a body with two surfaces occupying the same location (e.g., a crack)! Of course, any mathematical theory of elasticity excludes such irregular surface definitions in order to have a properly posed theory.

This chapter defines the degeneration of the BIE formulation and describes the modeling approaches for eliminating this limitation. Numerical results for the use of the BEM for fracture mechanics problems demonstrate both the accuracy and modeling efficiency of the method.

4.2 Degeneration of the BIE for Co-Planar Surfaces

The formulation of the BIE for elasticity was presented in Chapter 3. The formulation combined the analytical fundamental solution with the desired solution of the modeled problem, through Betti's reciprocal work theorem. The property of the fundamental solution led directly to internal solutions which identically satisfy internal equilibrium for displacements and stresses. These Somigliana identities are valid for very general boundaries, even non-smooth boundaries.

A simple description of a crack is two coplanar surfaces which are closed over some curve. Figure 4.1 serves to define some essential geometric terms for a three-dimensional crack. While the figure is for a crack lying in the x_1-x_3 plane, there is no inherent restriction that the crack be in a plane.

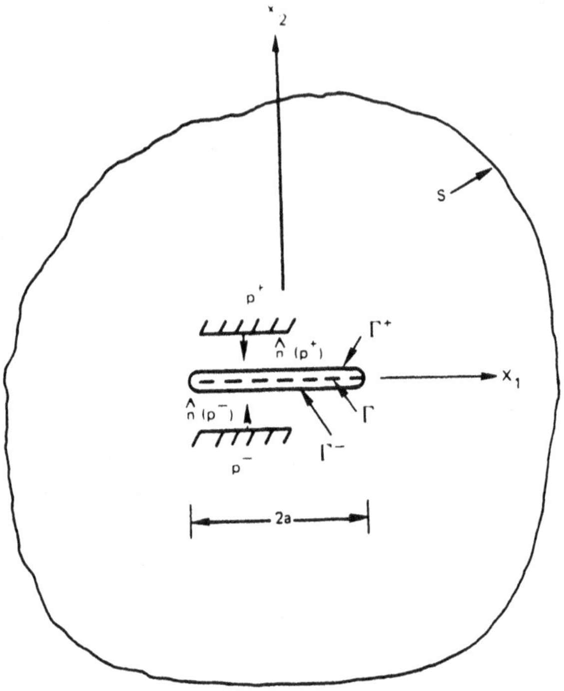

Figure 4.1. Flat crack modeling terms

The upper and lower crack surfaces are denoted Γ^+ and Γ^-, with normals $\underset{\sim}{n}(+)$ and $\underset{\sim}{n}(-)$, respectively. In the limit these two surfaces are defined as Γ (defined to have the normal $\underset{\sim}{n}(+)$), but with opposite direction normals. In two-dimensional plane strain problems the crack has length 2a and indefinite depth (x_3 direction).

Consider the Somigliana displacement identity of eq. (3.64), written for the crack surfaces Γ and for a regular surface S (which may intersect Γ), written as

$$
\begin{aligned}
u_i(p) = \int_S U_{ij}(p,Q)t_j(Q)dS + \int_{\Gamma^+} U_{ij}(p,Q)t_j(Q)\,dS \\
+ \int_{\Gamma^-} U_{ij}(p,Q)t_j(Q)\,dS \\
- \int_S T_{ij}(p,Q)U_j(Q)dS - \int_{\Gamma^+} T_{ij}(p,Q)u_j(Q)\,dS \\
- \int_{\Gamma^-} T_{ij}(p,Q)u_j(Q)\,dS
\end{aligned}
\tag{4.1}
$$

The fundamental solution kernels, $U_{ij}(p,Q)$ and $T_{ij}(p,Q)$, in eq. (4.1) for the crack surface Γ are symmetric and antisymmetric, respectively, for Γ^+ and Γ^-. Dropping the integral on the regular surface S, without loss of generality, eq. (4.1) may be written on the midsurface Γ as

$$u_i(p) = \int_\Gamma U_{ij}(p,Q)\,[t_j(Q^+) + t_j(Q^-)]\,dS \qquad (4.2)$$
$$-\int T_{ij}(p,Q)\,[u_j(Q^+) - u_j(Q^-)]\,dS$$

In the usual problems of fracture mechanics analysis, the tractions on the two crack surfaces are equal and opposite, such that

$$t_j(Q^+) = -\,t_j(Q^-) \qquad (4.3a)$$

The antisymmetry of eq. (4.3a) causes the square bracket traction term in eq. (4.2) to be identically zero for any equilibrated loading of the crack. If we define the difference in crack surface displacements as follows

$$\Delta u_j(Q) = u_j(Q^+) - u_j(Q^-) \qquad (4.3b)$$

then eq. (4.2) may be written as a new Somigliana identify for crack problems

$$u_i(p) = -\int_\Gamma T_{ij}(p,Q)\,\Delta u_j(Q)\,dS \qquad (4.4)$$

The BIE corresponding to the Somigliana identity for crack is obtained in the same manner as for eq. (3.74) as

$$u_i(P) - \Delta u_j(P)/2 + \int_{\langle\Gamma\rangle} T_{ij}(P,Q)\Delta u_j(Q)\,dS = 0 \qquad (4.5)$$

where $P(\tilde{x})$ is taken to be on Γ^+. The BIE that is obtained by letting $p(\tilde{x}) \to P(\tilde{x})$ on Γ^- (denoted P^-) may be found by remembering the normal in $\tilde{T}_{ij}(P,Q)$ is for Γ^+

$$u_i(P^-) + \Delta u_j(P)/2 + \int_{\langle\Gamma\rangle} T_{ij}(P,Q)\Delta u_j(Q)\,dS = 0 \qquad (4.6)$$

where the kernel in the integral is the same for both equations. Addition of eq. (4.5) to eq. (4.6) gives

$$\Sigma\, u_i(P)/2 + \int_{\langle\Gamma\rangle} T_{ij}(P,Q)\,\Delta u_j(Q)\,dS = 0 \qquad (4.7)$$

Equations (4.5-4.7) are identically the same and show that the usual BIE formulation for the fracture mechanics problem results in an integral eq. (4.7) in two unknowns, Σu_i and Δu_i. Further, the solution is not unique since any equilibrated applied tractions will give the same answer.

None of the cited BIE relations provides a means for the numerical solution of fracture mechanics problems. This was first noted by Cruse (1972) and expanded upon in two subsequent publications: Cruse (1974) and Cruse (1978). Figure 4.2 summarizes the early modeling approaches that were used to circumvent the degeneration of the BIE formulation for cracks.

Figure 4.2(a) gives a two-dimensional fracture mechanics problem with symmetry about the crack plane. Figure 4.2(b) illustrates the desired modeling approach, but with a degenerate crack surface model. The earliest successful treatment of the fracture mechanics problem by Cruse and Van Buren (1971) replaced the actual crack by an open notch, as in Fig. 4.2 (c). The open notch is of limited use, as the numerical conditioning improves with crack surface separation, but the crack approximation fidelity decreases. The accuracy of crack modeling for symmetric problems was seen to improve in Cruse (1974) by the modeling approach in Fig. 4.2(d). In this case the material surface ahead of the crack is replaced by a set of boundary elements with symmetric loading conditions.

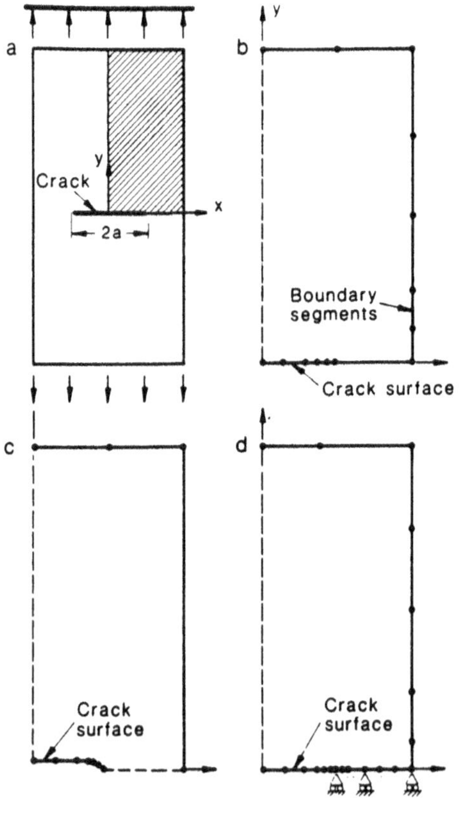

Figure 4.2. Flat crack modeling approaches

4.3 Multiregion BIE Applications

A substantial gain in BIE modeling was made by Lachat and Watson (1976) who wrote the first BIE code with a multiregion modeling capability. Consider two elastic bodies (α=1,2) joined along mutual surfaces $\hat{S}$. The BIE for each region is given by

$$C_{ij}^{\alpha} u_j^{\alpha} + \int_{\langle S^{\alpha}-\hat{S}\rangle} T_{ij}^{\alpha} (P,Q) u_j^{\alpha}(Q)\, dS + \int_{\langle \hat{S}\rangle} T_{ij}^{\alpha} (P,Q)\, u_j^{\alpha}(Q)\, dS$$
$$= \int_{\langle S^{\alpha}-\hat{S}\rangle} U_{ij}^{\alpha} (P,Q)\, t_j^{\alpha}(Q)\, dS + \int_{\langle \hat{S}\rangle} U_{ij}^{\alpha} (P,Q) t_j^{\alpha}(Q)\, dS \qquad (4.8)$$

In eq. (4.8), the superscript α is not summed.

The boundary conditions are those imposed separately on $S^{\alpha}-\hat{S}$, and the interface conditions on S

$$u_j^1 (Q) = u_j^2 (Q) \qquad Q \in S \qquad (4.9)$$
$$t_j^1(Q) = -t_j^2(Q)$$

The conditions in eq. (4.11) imply fully welded conditions. Sliding or friction boundary conditions may be imposed by using local boundary conditions (normal and tangential directions denoted by N,T)

$$u_j^{1N}(Q) = u_j^{2N}(Q)$$
$$t_j^{1T}(Q) = t_j^{2T}(Q) = 0 \qquad Q \in S \qquad (4.10)$$
$$t_j^{1N}(Q) = -\, t_j^{2N}(Q)$$

The multiregion modeling approach was first exploited for crack modeling by Blandford et al. (1981). Figure 4.3 reproduces their mesh for a two-dimensional, nonsymmetric crack geometry. Material is defined on both sides of the crack and is joined along arbitrary uncracked ligaments. The question of how to select the ligament is not insignificant, but the modeling approach is quite successful. More will be said on this in Section 4.4 on crack surface interpolations.

The system BIE model for multiregion problems is made up of non-zero terms for eq. (4.8) for each material, but is not a full system of equations. Significant blocks of zero terms occur after imposing boundary conditions such as (4.9) and (4.10), and should be taken into account by the equation solution algorithm. The paper by Lachat and Watson (1976) should be consulted as well as numerous subsequent papers by various authors.

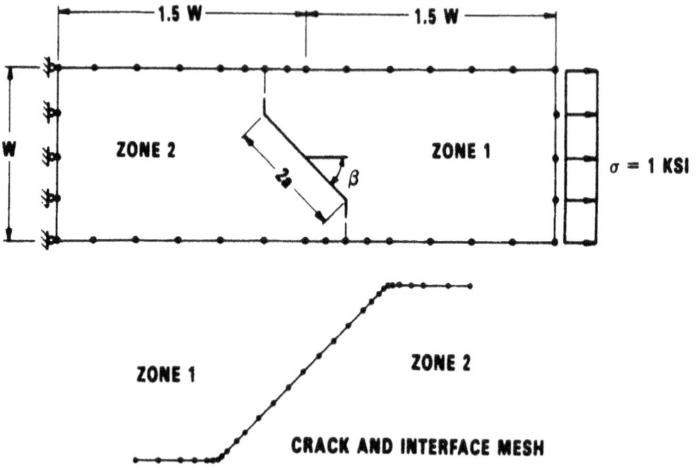

Figure 4.3. Multiregion crack modeling

4.4 Strain Energy Based Crack Tip Modeling

The direct application of standard BIE analysis of crack geometries has been shown to involve modeling of entire crack planes--the crack and the ligament. The crack opening displacements for points asymptotically close to the crack tip were shown in Chapter 2 to be related to the stress intensity factor K in the form

$$K_\beta = \lim_{\varepsilon \to 0} \frac{H\sqrt{2\pi}}{4} \frac{\delta_\beta(\varepsilon)}{\sqrt{\varepsilon}} \qquad (\beta = I, II, III) \tag{4.11}$$

In eq. (4.13), β refers to the mode of crack tip response or loading, $H = E/(1-\nu^2)$ in three dimensions and in plane strain, ε is the distance from the crack tip, and δ_β is the crack tip motion.

As in finite element modeling of crack problems, crack tip motion, $\delta(\varepsilon)$, is the most reliable measure or estimate for the stress intensity factor. Equation (4.11) has been used to obtain K by graphical or a numerical limiting process, as in Cruse (1974). Another very successful and more accurate method uses $\delta(\varepsilon)$ in an energy context, as developed by Cruse and Meyers (1977).

As discussed in Chapter 2, the stress intensity factor is directly related to the strain energy release rate (**G**) for crack extension

$$K^2 = HG \tag{4.12}$$

for two-dimensional problems. The strain energy release rate may be numerically estimated by a finite difference of the strain energy

$$G = \frac{U_2 - U_1}{A_2 - A_1} \tag{4.13}$$

where the subscripts 1,2 refer to the original and the larger crack size. The strain energy U for the elastic body is equal to the work done by the boundary fractions, and is easily and directly computed from the BIE solution. The sign changes in eq. (4.13) from eq. (2.1) due to the difference in fixed versus non-fixed boundary conditions remote from the crack.

For three-dimensional problems, the stress intensity factor is a function of position along the crack front. The three-dimensional solution for an elliptical crack is summarized in eq. (2.16), where it is shown that, for uniform symmetrical loading of the crack,

$$K_I = \frac{\sigma\sqrt{\pi a}}{E(k)} \{\sin^2\phi + (a/b)^2 \cos^2\phi\}^{1/4} \qquad a<b \tag{4.14}$$

In eq. (4.16), a and b are elliptical semi-minor and semi-major axis lengths, σ is the uniform applied stress, and ϕ is the elliptical angle as shown in Fig. 2.3. As will be demonstrated in Chapter 6, the crack opening displacements for the three-dimensional problem near the crack tip are given by combining the stress solution in eq. (4.14) with the asymptotic crack opening equation obtained by inverting eq. (4.11).

For problems with varying K(s), along the crack front parameter, s, the strain energy release rate for a small crack front change was shown by Cruse and Meyers (1977) to be given by

$$G = \frac{1}{H\Delta A} \int_{\Delta A} K^2(s)\ \delta\ell(s)\ ds \tag{4.15}$$

where $\delta\ell(s)$ is the small distance of crack extension, measured normal to the original crack front; ΔA is the new crack surface area; and H is $E/(1-\nu^2)$. As done by Cruse and Besuner (1975) and by Cruse and Meyers

(1977), a weighted root-mean-square stress intensity factor, $\bar{K}$, is obtained by combining eqs. (4.12) and (4.16)

$$\bar{K}^2 = GH = \frac{1}{\Delta A} \int_{\Delta A} K^2(s)\, \delta\ell(s)\, ds \qquad (4.16)$$

where **G** is computed by eq. (4.13).

Figure 4.4 illustrates two approaches to the application of eq. (4.16) to three-dimensional crack front modeling. In the first half of the figure, all the crack front nodes are perturbed (the figure exaggerates the pertubation considerably), while in the second half of the figure only one node is perturbed. The first approach was used by Cruse and Besuner (1975) to reduce elliptical cracks to two degrees of freedom by allowing elliptical cracks to be distorted into other elliptical cracks, changing only the size parameters (a,b). This approach smears the K-distribution out some along the crack front. The local pertubation approach provides a good estimate of the local value of K(s), and is not limited to elliptical cracks; of course, the number of degrees of freedom is greatly increased.

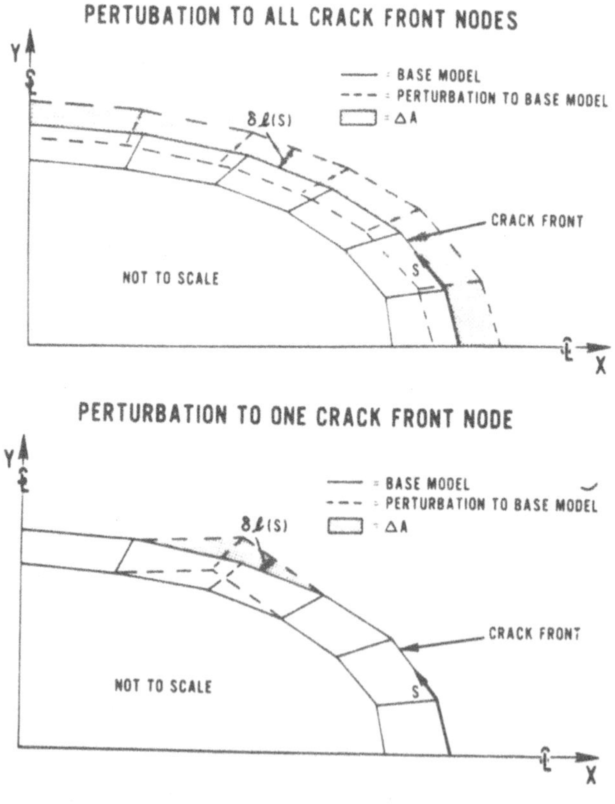

Figure 4.4. Crack front perturbation approaches

Table 4.1 compares local stress intensity factor values computed for a linear variation BEM model of the crack, given by Cruse and Meyers (1977) for a 2:1 aspect ratio buried crack subject to simple tension (see eq. (4.14). The BEM mesh for this problem is given in Figs. 4.5 and 4.6, where symmetry conditions are imposed to model the buried crack. The data in the first column are K-estimates taken from the crack opening at the near-tip nodes, using eq. (4.11). The errors are about 8% at each of the crack front locations.

Table 4.1. Comparison of method for opening values of K for 2:1 buried ellipse

ϕ^* (1)	K^{cod}/K^{exact} [b,c] (2)	K^g/K^{exact} [b,d] (4)
0	0.916	0.951
15	0.923	0.958
30	0.023	0.959
45	0.924	0.960
60	0.925	0.961
75	0.929	0.965
90	0.937	0.974

*elliptical angle from ellipse major axis.

[b]K^{exact} = exact value of K.

[c]K^{cod} = values of K using crack opening displacements.

K^G = values of K using strain energy rate procedure.

A local distribution for K(s) was obtained by normalizing the crack opening values to a reference node such that

$$K(s)/K_{ref} = [\delta(s)/\delta_{ref}] \times [\varepsilon_{ref}/\varepsilon(s)]^{1/2} = \delta^*(s)/\sqrt{\varepsilon^*(s)} \tag{4.17}$$

Combining eq. (4.17) with (4.16) we obtain

$$GH = \frac{1}{\Delta A}\int_{\Delta A} K^2_{ref}\left[\frac{\delta^*(s)}{\sqrt{\varepsilon^*(s)}}\right]^2 \delta\ell(s)\, ds \tag{4.18}$$

The reference value of K can be obtained by rearranging the terms in eq. (4.20) to obtain

$$K^2_{ref} = \frac{1}{\Delta A}\int_{\Delta A} \frac{[\delta^*(s)]^2}{\varepsilon^*(s)}\,\frac{\delta\ell(s)ds}{H} \tag{4.19}$$

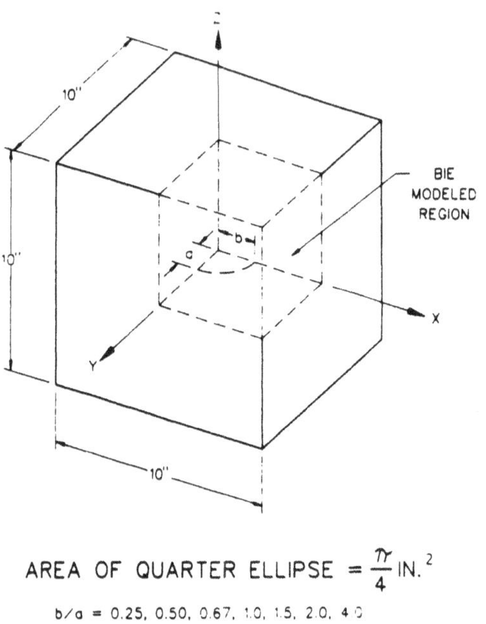

Figure 4.5. Cube containing buried elliptical crack

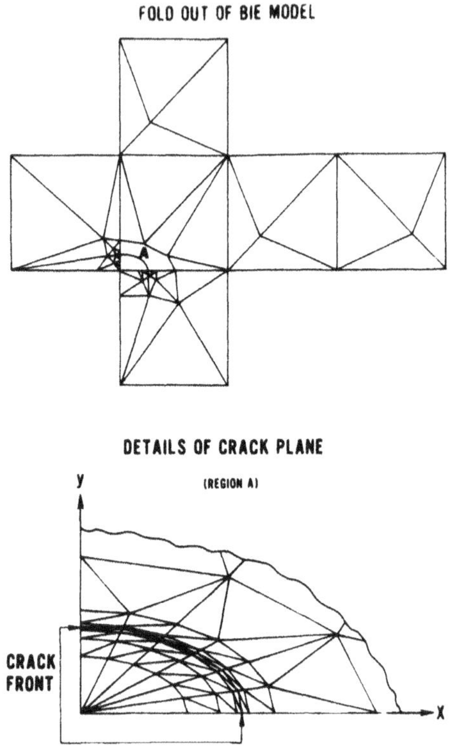

Figure 4.6. BIE model for buried elliptical crack in cube

The values of $\delta^*(s)$, $\varepsilon^*(s)$ are obtained from the BEM model of the unperturbed crack. G is estimated from the change in strain energy computed for a given change in crack shape given by $\delta\ell(s)$. Values of K(s) then come from eq. (4.17).

The second column of data in Table 4.1 shows the values of K(s) obtained in this manner by taking a small, constant distribution, $\delta(s)$ = constant, and using strain energy calculation as in eq. (4.19). The error is seen to reduce by about 1/2 to 4%, for this BEM model. The results obtained by a local perturbation of the crack front were within ±1.5% of these latter values.

Thus, the energy-based, crack extension method is seen to obtain good accuracy for three-dimensional problems, at the cost of an additional BEM analysis. The next section shows a substantial gain in accuracy through the use of improved boundary data interpolation near the crack tip.

4.5 Crack Surface Interpolations

The data just reported for three-dimensional cracks used standard, linear interpolation of the boundary data; no special treatment of the crack opening shape near the tip was used. Finite element models (FEM) of cracks demonstrated the need for special crack tip modeling in order to achieve even reasonable numerical accuracy. Barsoum (1976) exploited a special feature of the quadratic shape functions available in most FEM codes to substantially improve FEM accuracy for fracture mechanics problems.

What Barsoum found was that the strain field obtained by differentiation of these displacement shape functions could be singular for a range of locations of the side nodes, for element sides focusing at the crack tip. When the side nodes were placed at one-quarter of the length of the side, away from the tip, a square-root-distance singularity in strain (and hence stress) was obtained.

While the results of Barsoum applied to the continuum modeling of the FEM, it is possible to exploit a similar approach for BEM models of crack surfaces. Consider the rectangular boundary element shown in Fig. 4.7. The displacement field for this element is to be interpolated using the eight-noded, isoparametric shape functions

$$
\begin{aligned}
N_1(\xi) &= (\xi+1)\,(\eta+1)\,(\xi+\eta-1)/4 \\
N_2(\xi) &= (\xi-1)\,(\eta+1)\,(\xi-\eta+1)/4 \\
N_3(\xi) &= (1-\xi)\,(\eta-1)\,(\xi+\eta+1)/4 \\
N_4(\xi) &= (\xi+1)\,(\eta-1)\,(\eta-\xi+1)/4 \\
N_5(\xi) &= (\xi+1)\,(1-\eta^2)/2 \\
N_6(\xi) &= (\eta+1)\,(1-\xi^2)/2 \\
N_7(\xi) &= (\xi-1)\,(\eta^2-1)/2 \\
N_8(\xi) &= (1-\eta)\,(1-\xi^2)/2
\end{aligned}
\qquad (4.20)
$$

Recalling from eq. (3.81) that the geometry model uses the same shape functions, we find that the geometry coordinate (y) is given for this element as

$$y = L(\eta-1)^2\,/4 \qquad (4.21)$$

where η is the second intrinsic (or unit square mapping) coordinate. Rewriting eq. (4.21) we obtain

$$\eta = 1+2(y/L)^{1/2} \tag{4.22}$$

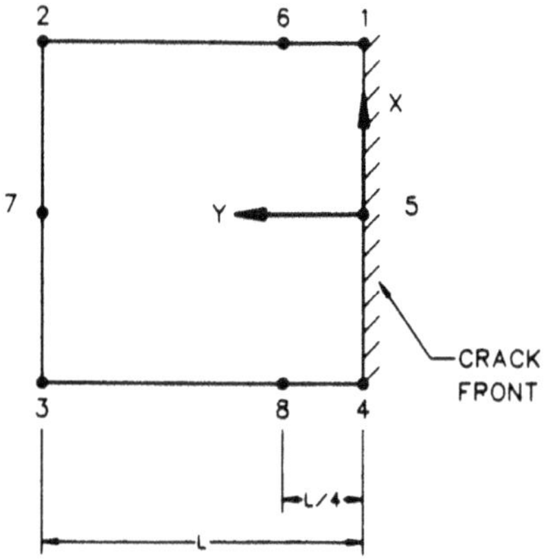

Figure 4.7. Quarter-point modeling approach

Since the crack surface motion is also written using eq. (3.81) in terms of linear and quadratic terms in η, the quarter-point geometry mapping produces crack surface motions of the general form

$$u \sim A + B(y/L)^{1/2} + C(y/L) + \cdots \tag{4.23}$$

where A=0 for the crack tip problem.

Equation (4.23) contains all of the essential crack surface motion terms obtained by a Taylor series expansion of the exact solution. The form of eq. (4.23) with the correct square-root behavior applies only if the element is rectangular, as shown in Fig. 4.7. When the crack front is curved, the quarter-point modeling system does not achieve the correct square-root power term. However, for elements with large radius of curvature of "L" ratios, the power is very close to square-root.

The modeling scheme discussed in Section 4.2 and shown in Fig. 4.2(d) also requires that the stresses (tractions) ahead of the crack tip be interpolated. The boundary tractions, of course, are modeled using the same shape functions as the displacements, as given by Eq. (3.81). Unlike the FEM approach, there is no direct way to induce singular tractions in eq. (3.81) through simple manipulation of the side-node locations.

Cruse and Wilson (1978) first introduced the notion of singular crack surface tractions by an explicit singular term. This was exploited extensively by Heliot et al. (1979) and his co-workers for a variety of surface crack problems.

The essential step is to take the tractions to be given by

$$t_i(Q) = \hat{t}_i(Q)\ (-y/L)^{1/2} \qquad (4.24)$$

in the surface integral in the BIE, for the rectangular problem shown in Fig. 4.7. The negative-y is depicted in order to convey the notion that we are now dealing with points ahead of the crack (y<0) in Fig. 4.7. The isoparametric shape functions are then used to interpolate $\hat{t}_i(Q)$ in eq. (4.24). Combination of eq. (4.24) with eq. (4.23) results in a near-tip stress expansion that contains the square-root singular stress term, a constant stress term, and higher order terms. These accurately reflect the full exact solution near the crack tip.

Table 4.2 summarizes the results for crack tip stress intensity factor accuracy, for the problem of a circular bar with a circular buried crack. The BEM mesh is shown in Fig. 4.8. The problem is simple tension transverse to the crack plane. Mode I stress intensity factors are calculated using the displacement relations from eq. (4.13) for the first node behind the crack tip.

Table 4.2. Crack singularity model results

Method	K_I/K_I(exact)
Standard BEM	1.10
Quarter Points Only	1.04
Quarter Points with Traction Singularity	1.01

The second quarter-point model used quarter points for elements behind <u>and ahead</u> of the crack front. The results for this approach are better than for just the displacement element because the quarter-point approach puts more Gaussian integration points near the crack tip; this is especially needed because of the stress gradient at the tip.

The use of the quarter-point modeling with the traction singular element is seen to produce very good accuracy, even for the relatively crude mesh used. The numerical results for $\hat{t}_i$ in eq. (4.24) for the crack tip nodes are not as accurate an estimate of the stress intensity factor as is the crack opening displacement. Improved accuracy in $\hat{t}_i$ is seen as the order of the Gaussian integration is increased, showing the impact on accuracy of the singular weight function in eq. (4.24).

Figure 4.9 compares the BEM data from the mesh in Fig. 4.10 to the FEM results of Newman and Raju (1979). The agreement, using the traction singular mesh, is quite satisfactory. A comparison of the BEM mesh with the FEM mesh of Newman and Raju (1979) will convince all but true FEM die-hards of the modeling efficiency of the BEM over the FEM for crack problems.

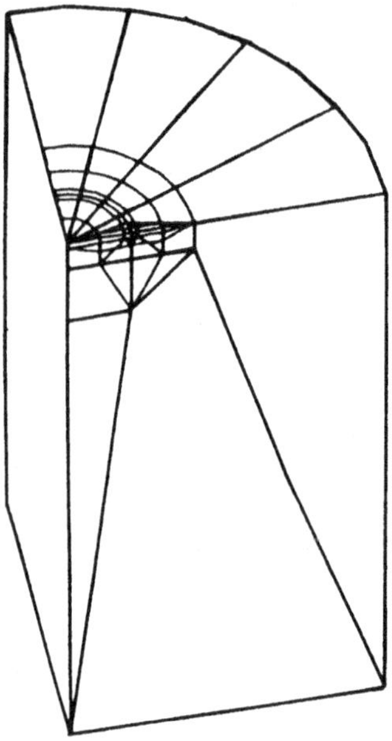

Figure 4.8. Circular buried crack geometry (1/8 symmetry)

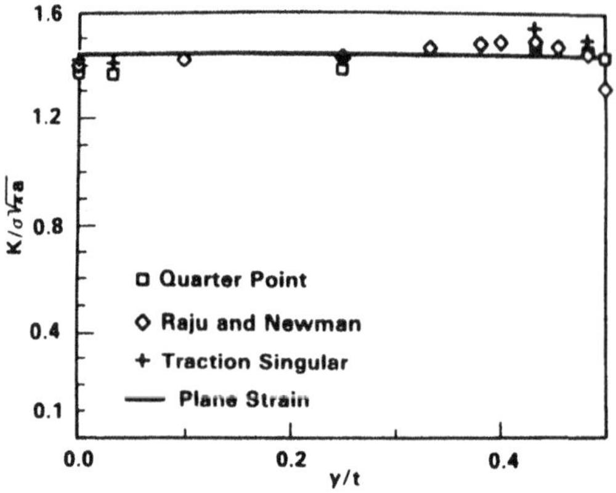

Figure 4.9. Stress intensity factor comparison

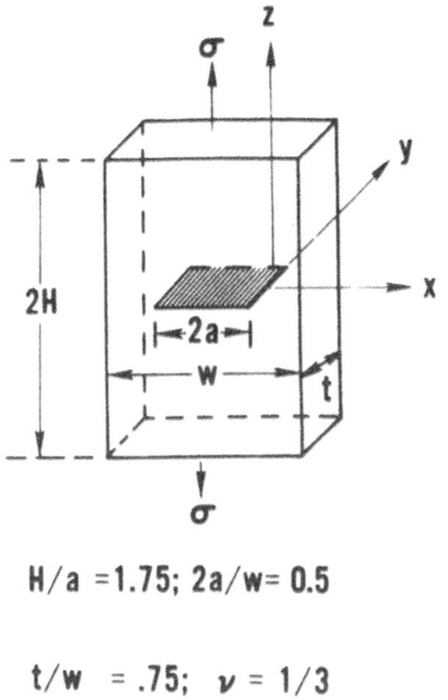

(a)

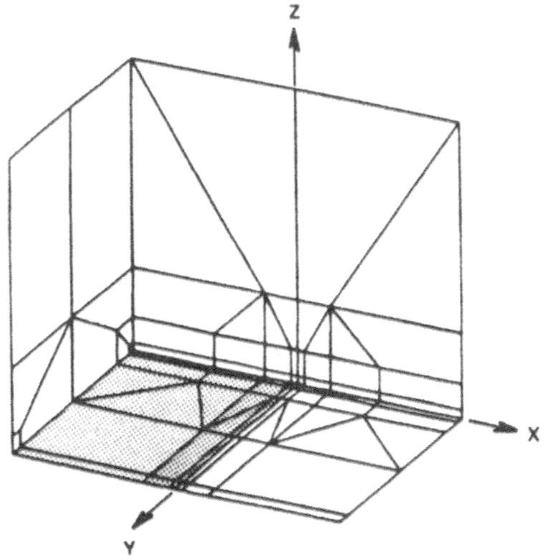

(b)

Figure 4.10. Center cracked test specimen geometry and BEM mesh

(1969) T.A. Cruse, Numerical Solutions in Three-Dimensional Elastostatics, International Journal of Solids and Structures, 5, 1259-1274.

(1971) T.A. Cruse and W. Van Buren, Three-Dimensional Elastic Stress Analysis of a Fracture Specimen with an Edge Crack, International Journal of Fracture Mechanics, 7, no. 1, 1-15.

(1972) T.A. Cruse, Numerical Evaluation of Elastic Stress Intensity Factors by the Boundary-Integral Equation Method, The Surface Crack: Physical Problems and Computational Solutions, (J.L, Swedlow, ed.) American Society of Mechanical Engineers, 153-170.

(1974) T.A. Cruse, An Improved Boundary-Integral Equation Method for Three Dimensional Elastic Stress Analysis, Computers and Structures, 4, 741-754.

(1975) T.A. Cruse and P.M. Besuner, Residual Life Prediction for Surface Cracks in Complex Structural Details, Journal of Aircraft, 12, 369-375.

(1976) R.S. Barsoum, On the Use of Isoparametric Finite Elements in Linear Fracture Mechanics, International Journal of Numerical Methods in Engineering, 10, 25-37.

(1976) J.C. Lachat and J.O. Watson, Effective Numerical Treatment of Boundary Integral Equations: A Formulation for Three-Dimensional Elastostatics, International Journal for Numerical Methods in Engineering, 10, 991-1005.

(1977) T.A. Cruse and G.J. Meyers, Three-Dimensional Fracture Mechanics Analysis, Journal of the Structural Division, American Society of Civil Engineers, 103, 309-320.

(1978) T.A. Cruse, Two Dimensional BIE Fracture Mechanics Analysis, Applied Mathematical Modeling, 2, 287-293.

(1978) T.A. Cruse and R.B. Wilson, Advanced Applications of Boundary-integral Equation Methods, Nuclear Engineering and Design, 46, 223-234.

(1979) J. Heliot, R.C. Labbens, and A. Pellissier-Tanon, Semi-Elliptical Cracks in a Cylinder Subjected to Stress Gradients, Fracture Mechanics (C.W. Smith, ed.), American Society for Testing and Materials STP 677, 341-364.

(1979) J.C. Newman and I.S. Raju, Analysis of Surface Cracks in Finite Plates Under Tension or Bending Loads, NASA TP-1578.

(1981) G.E. Blandford, A.R. Ingraffea, and J.A. Liggett, Two-Dimensional Stress Intensity Factor Computations Using the Boundary Element Method, International Journal for Numerical Methods in Engineering, 17, 387-404.

5.0 Green's Function Formulation in Two Dimensions

5.1 Introduction

The previous chapter outlined some of the major issues in the direct application of the BIE to modeling crack problems. The fact that the BIE ceases to be meaningful for problems with two coplanar crack surfaces forces the use of multiregion or symmetric crack plane models. Both of these procedures require modeling of some surface connecting the crack tip surface(s) with other surfaces. In addition, special interpolation methods are required to achieve good engineering accuracy at reasonable cost.

Two-dimensional elasticity formulations have access to the powerful tools of complex variable functions for the analytical treatment of a wide variety of special boundary shapes. These tools have provided the means of formulation for a very efficient and accurate BIE method for certain classes of two-dimensional elastic fracture mechanics problems.

In the current formulation, the fundamental solution for the anisotropic elasticity problem (see Section 3.4) is modified to include an exact solution for a stress-free crack as shown in Fig. 5.1. The surface L(x) is flat and extends from x = -a to x = +a. The use of special geometric features for which the boundary conditions are treated exactly is referred to in the mathematics of integral equations as Green's functions for that geometry; see for example, Greenberg (1971).

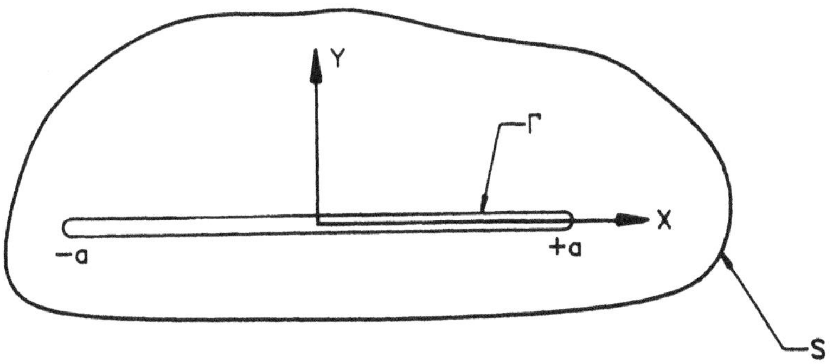

Figure 5.1. Two-dimensional crack geometry

We shall see in the subsequent sections of this Chapter how beneficial Green's functions can be to BIE formulations. Special Green's functions for geometries other than the single crack, such as the elliptical hole, curved crack, and multiple crack problems may be derived in much the same way as for the crack.

5.2 Formulation of the Anisotropic Green's Function

The fundamental solution corresponding to the point load in the two-dimensional anisotropic plane was developed in Section 3.4. Let this function be given by

$$\Phi^{o}_{jk} = A_{jk}\log(z_k - c_k) \tag{5.1}$$

where the anisotropic coefficients A_{jk} are defined in eq. (3.45), where $z_k = x + \mu_k y$, and μ_k is defined by eq. (3.35), and where c_k is the point of load application.

The Green's function formulation is based on combining the point-load solution Φ^{o}_{jk} with a second solution $\Phi^{*}_{jk}(z_k)$ such that the crack boundary conditions are identically satisfied. For the current development of a single, straight crack along the x-axis, the boundary conditions from eq. (3.40) may be written

$$\sigma_y + i\tau_{xy} = \beta(x) \qquad -a \le x \le a \qquad i = \sqrt{-1} \tag{5.2}$$

where $\beta(x)$ is the applied stress on the crack surface $L(x)$. Using the complex variable stress functions, the general form of eq. (5.2) may be written, remembering that $\Phi_{jk} = dF_{jk}/dz_k$, as

$$\sum_{k=1}^{2} [d\Phi^{*}_{jk}/dz_k + d\bar{\Phi}^{*}_{jk}/d\bar{z}_k - i\mu_k d\Phi^{*}_{jk}/dz_k - i\bar{\mu}_k d\bar{\Phi}^{*}_{jk}/d\bar{z}_k] = \beta(z) \tag{5.3}$$

where $i = \sqrt{-1}$ and the subscript -j refers to the direction of the point load. Eq. (5.3) may be simplified by letting $\gamma_k = 1 - i\mu_k$ and $\delta_k = 1 + i\mu_k$, such that

$$\sum_{k=1}^{2} [\gamma_k\, d\Phi^{*}_{jk}/dz_k + \bar{\delta}_k d\bar{\Phi}^{*}_{jk}/d\bar{z}_k] = \beta_j(z) \tag{5.4}$$

or

$$h_j(z) + k_j(z) = \beta_j(z) \tag{5.5}$$

where

$$h_j(z) = \sum_{k=1}^{2} [\gamma_k \, d\Phi^*_{jk}/dz_k]$$

$$\text{and} \quad k_j(z) = \sum_{k=1}^{2} [\bar{\delta}_k \, d\bar{\Phi}^*_{jk}/d\bar{z}_k] \tag{5.6}$$

The boundary conditions are then imposed by evaluation of eq. (5.4) for both the upper and lower crack surfaces, assuming that $\beta_j^+(x)$ and $\beta_j^-(x)$ are equal and opposite (equilibrated) stresses. The boundary conditions become

$$h_j^+(x) + k_j^-(x) = \beta_j(x)$$

$$-a \leq x \leq a \tag{5.7}$$

$$h_j^-(x) + k_j^+(x) = \beta_j(x)$$

The reversal in the sign symbols in eq. (5.7) occurs due to the conjugation relation on z_k in eq. (5.6); that is, an upper surface point z_k corresponds to a lower surface point $\bar{z}_k$.

The general form of the problem in eq. (5.7) is a Hilbert problem, as discussed for example, in the book by England (1971), and discussed in depth for this problem by Snyder (1973). The solution to the Hilbert problem for eq. (5.7) is given as

$$h_j(z) + k_j(z) = \frac{\chi(z)}{\pi i} \int_L \frac{\beta_j(t)\, dt}{\chi^+(t)(t-z)} + P_j(z)\, \chi(z) \tag{5.8}$$

and

$$h_j(z) - k_j(z) = Q_j(z) \tag{5.9}$$

where $P_j(z)$, $Q_j(z)$ are positive power polynomials. For the single crack, $L(x)$, from $x = -a$ to $x = +a$, $\chi(z)$ is given by

$$\chi(z) = (z^2-a^2)^{-1/2} \tag{5.10}$$

The polynomials $P_j(z)$, $Q_j(z)$ may be determined from the stress conditions at infinity and, to within a rigid body term, are zero.

Setting $P_j(x) = Q_j(z) = 0$ in eq. (5.8) and (5.9) results in the relation

$$h_j(z) = k_j(z) = \frac{1}{2\pi\sqrt{z^2-a^2}} \int_{-a}^{+a} \frac{\sqrt{[a^2-t^2]}\ \beta_j(t)\,dt}{t-z} \tag{5.11}$$

Solving eq. (5.6) for the derivatives of the stress functions, we find that

$$\frac{d\Phi^*_{j1}}{dz_1} = \frac{\delta_2 h_j(z) - \overline{\gamma_2 k_j(\bar{z})}}{2i(\mu_2-\mu_1)} = \frac{\delta_2 h_j(z_1) - \overline{\gamma_2 k_j(\bar{z}_2)}}{2i(\mu_2-\mu_1)} \tag{5.12}$$

$$\frac{d\Phi^*_{j2}}{dz_2} = \frac{-\delta_1 h_j(z) + \overline{\gamma_1 k_j(\bar{z})}}{2i(\mu_2-\mu_1)} = \frac{-\delta_1 h_j(z_2) - \overline{\gamma_1 k_j(\bar{z}_2)}}{2i(\mu_2-\mu_1)} \tag{5.13}$$

That eq. (5.12) and (5.13) are functions of z_1, z_2 respectively, follows from the analytical nature of the functions and that $z_1 = z_2 = z$ on $L(x)$.

For the point-load stress functions in eq. (5.1) and the definitions in Section 3.10, it can be shown that the stresses from the point load are on the line of the crack location given by

$$(\sigma_y + i\tau_{xy})_j = -\beta_j(x) = \sum_{k=1}^{2} \left[\frac{\gamma_k A_{jk}}{x-c_k} + \frac{\bar{\delta}_k \bar{A}_{jk}}{x-\bar{c}_k}\right] \tag{5.14}$$

where $c_k = x_o + \mu_k y_o$, the location of the point load. Then, from eq. (5.11), the necessary auxiliary solution to create a stress-free crack is given by

$$h_j(z) = k_j(z) = -\frac{1}{2\pi\sqrt{(z^2-a^2)}} \int_{-a}^{a} \frac{\sqrt{(a^2-t^2)}}{t-z} \left(\sum_{k=1}^{2} \left[\frac{\gamma_k A_{jk}}{t-c_k} + \frac{\bar{\delta}_k \bar{A}_{jk}}{t-\bar{c}_k}\right]\right) dt \tag{5.15}$$

The integrals may be computed to obtain

$$h_j(z) = k_j(z) = -\frac{1}{2\pi\sqrt{(z^2-a^2)}}\left[\sum_{k=1}^{2}\left(\frac{\gamma_k A_{jk}}{z-c_k}\left[I(z)-I(c_k)\right]\right.\right. \qquad (5.16)$$

$$\left.\left.+\frac{\bar{\delta}_k \bar{A}_{jk}}{z-\bar{c}_k}\left[I(z)-I(\bar{c}_k)\right]\right)\right]$$

where

$$I(z) = \int_{-a}^{a} \frac{\sqrt{(a^2-t^2)}\,dt}{t-z} = \pi[\sqrt{(z^2-a^2)} - z] \qquad (5.17)$$

The results in eq. (5.16) can be substituted into eq. (5.12) and (5.13) and integrated to obtain the stress function for the stress-free crack. Combined with the point-load stress function, eq. (3.44), we obtain, following Snyder (1973)

$$\Phi_{jk}(z_k) = -\frac{1}{2\pi(\mu_\ell-\mu_k)}\left[(\mu_\ell-\mu_k)A_{jk}J(z_k,c_k)\right.$$

$$\left.+(\mu_\ell-\bar{\mu}_\ell)\,\bar{A}_{j\ell}\,J(z_k,\bar{c}_\ell)+(\mu_\ell-\bar{\mu}_k)\,\bar{A}_{jk}\,J(z_k,\bar{c}_k)\right] \qquad (5.18)$$

$$+A_{jk}\log(z_k-c_k)$$

where $\ell = 3-k$, and where $J(z,c)$ is given as the integral

$$J(z,c) = \int \frac{I(z) - I(c)}{(z-c)\sqrt{(z^2-a^2)}}\,dz \qquad (5.19)$$

Snyder's (1973) results, reported by Snyder and Cruse (1975), used an integral result for eq. (5.19) that was correct, to within a complex constant of integration. In order to avoid violating the branch cut of the crack line, Cruse (1978) showed the correct integral of eq. (5.19) was given by

$$J(z,c) = \pi\log\left\{\frac{\sqrt{(z^2-a^2)}\,\sqrt{(c^2-a^2)} + cz-a^2}{[z+\sqrt{(z^2-a^2)}]\,[c+\sqrt{(c^2-a^2)}]}\right\} \qquad (5.20)$$

The fundamental displacement Green's function is given, for each component of the point load (denoted by the j-index), by substitution of eq. (5.18) into eq. (3.41)

$$U_{ji}(p,q;a) = 2\mathrm{Re}\left[\sum_{k=1}^{2} P_{ik}\Phi_{jk}(z_k,c_k)\right] \tag{5.21}$$

where the matrix for P_{ik} is given by

$$\{P_{ik}\} = \begin{Bmatrix} \beta_{11}\mu_k^2 + \beta_{12} - \beta_{16}\mu_k \\ \beta_{22}/\mu_k + \beta_{12}\mu_k - \beta_{26} \end{Bmatrix} \tag{5.22}$$

The explicit functional dependence of the Green's function fundamental solution in eq. (5.21) on the crack length is emphasized by the notation in the parenthetical terms.

The fundamental traction Green's function corresponding to the displacement solution (5.21) is given by the usual combination of stress terms and the unit outward normal vector n_i. After some manipulation, the general form is given by

$$T_{ji}(p,q;a) = 2\mathrm{Re}\left[\sum_{k=1}^{2} Q_{ik}(\mu_k n_1 - n_2)d\Phi_k(z_k,c_k)/dz_k\right] \tag{5.23}$$

where the matrix Q_{ik} is given by

$$\{Q_{ik}\} = \begin{Bmatrix} \mu_k \\ 1 \end{Bmatrix} \tag{5.24}$$

Letting $B_k = (\mu_k n_1 - n_2)$ the expansion of terms for the traction Green's function is given by

$$T_{ji}(p,q;a) = 2\mathrm{Re}\left\{\sum_{k=1}^{2}\left[Q_{ik}B_k A_{jk}\left(\frac{1}{z_k-c_k}\right) - \frac{Q_{ik}B_k}{2\pi(\mu_\ell-\mu_k)}\left\{(\mu_\ell-\mu_k)\,A_{jk}\frac{\partial J(z_k,c_k)}{\partial z_k}\right.\right.\right.$$

$$\left.\left.\left. + (\mu_\ell-\bar{\mu}_\ell)\bar{A}_{j\ell}\frac{\partial J(z_k,\bar{c}_\ell)}{\partial z_k} + (\mu_\ell-\bar{\mu}_k)\,\bar{A}_{jk}\frac{\partial J(z_k,\bar{c}_k)}{\partial z_k}\right\}\right]\right\} \tag{5.25}$$

The first term in eq. (5.25) is the simple pole for the two-dimensional point-load solution. All of the remaining terms assure that the crack is stress free for any location of p,q, including on the crack surface. These additional terms contain the information on crack size, location, and orientation.

5.3 Somigliana Identities for the Anisotropic Green's Function Formulation

The formulation of the Somigliana identities given in Section 3.6 applies to the Green's function formulation of this Chapter. The problem currently posed is anisotropic, linear elastic such that eq. (3.59) applies throughout the volume.

Consider a finite geometry bounded by the regular surface S and containing the crack denoted Γ. The crack surface and the regular surface may intersect at any finite number of points not at the crack tips. The Green's function formulation in Section 5.2 assumes the origin to be at the center of the crack and that the crack is along the +/- x-axis. Application of an origin shift and axis rotation to the Green's function formulation is simply accomplished and nothing further will be presented on this modification.

Following the notation of eq. (3.60), the fundamental solution is denoted by the superscript asterisk, and as previously shown

$$\int_V \sigma_{ij}\, u^*_{i,j}\, dV = \int_{S+\Gamma} t_i u^*_i\, dS \tag{5.30}$$

where the body force for the posed problem is taken to be zero.

Further, assume that the problem to be solved has zero applied traction on the crack surface Γ; i.e.,

$$t_i(Q) \equiv 0 \qquad Q \in \Gamma \tag{5.31}$$

This boundary condition corresponds to the assumed loading in the Green's function formulation of Section 5.2.

Application of the divergence theorem as in eq. (3.62) results in the following identity

$$\int_V \sigma^*_{ij}\, u_{i,j}\, dV = \int_{S+\Gamma} t^*_i\, u_i\, dS + u_i e_i \tag{5.32}$$

The free term exists as before due to the fundamental solution inverse r(p,q) term in eq. (5.25). Equating the two volume integrals from eq. (5.31) and (5.32), and recognizing zero crack surface tractions, we obtain the Somigliana displacement identity

$$u_i(p) = \int_S U_{ij}(p,Q;a)t_j(Q)\,dS - \int_S T_{ij}(p,Q;a)u_j(Q)\,dS \tag{5.33}$$

We thus see that the displacement solution to the fracture mechanics problem does not include any integrals on the crack surface.

If, in fact, the applied traction on the crack is non-zero, the Green's function boundary conditions, eq. (5.14), are augmented by the addition of the physical tractions. The stress function for the added tractions may be obtained from integration of eq. (5.15) for the augmented traction term. The resulting augmented solution displacement field is the infinite plate solution, of course, for the applied loading. It can be shown that application of the reciprocal work theorem will result in cancellation of the crack surface loading terms from both sides of the equality. While the form of the Somigliana identity is equivalent to eq. (5.33), the Green's function displacement kernel contains the effect of the applied crack surface traction. This is functionally equivalent to application of superposition to eliminate the crack surface loading term. Thus, without loss of generality, the zero traction form of the Somigliana identity is complete.

The displacement identity in eq. (5.33) is valid for any points $p(\underset{\sim}{x})$ including $p(\underset{\sim}{x})\varepsilon\Gamma$. Some care must be exercised in evaluating the kernels, but upper and lower crack surface limiting forms exist. As a practical matter, the application of eq. (5.33) for crack surface points is most easily made <u>after</u> discretizing the integrals.

The displacement identity may be differentiated at $p(\underset{\sim}{x})$ as in eq. (3.65). Taking the symmetric part of the gradient operator on eq. (5.33) we obtain the interior strain identity

$$\varepsilon_{ij}(p) = \int_S U^{\varepsilon}_{ijk}(p,Q;a)t_k(Q)dS - \int_S T^{\varepsilon}_{ijk}(p,Q;a)u_k(Q)dS \tag{5.34}$$

Differentiation of the displacement Green's function terms due to the crack, to obtain eq. (5.34), is straightforward, although cumbersome due to the form of such terms as (5.20) in eq. (5.21) and eq. (5.25). It can be shown that each of these terms is of the form

$$F(z,c;a)/\surd(z^2-a^2) \tag{5.35}$$

where F(z,c;a) is a regular function.

The presence of this inverse singular term is, of course, fully expected from eq. (5.10) in the formulation of the Green's function for the crack problem. The form of eq. (5.35) for points near either crack tip, $z = \pm(a \mp \Delta a)$ for $\Delta a/a \ll 1$, leads to the usual $\sqrt{R}$ singularity in strains where $R=\Delta a$, the distance from the crack tip. As shown in Chapter 2, the regular coefficient of this singular term is proportional to the strain intensity factor K^ε

$$K^\varepsilon = \lim_{\Delta a \to o} \left(\sqrt{2\pi\Delta a}\ \varepsilon[\pm(a \mp \Delta a)]\right) \qquad (5.36a)$$

The formal limit in eq. (5.36a) may be taken on eq. (5.35) with the symbolic result

$$K^\varepsilon = \pm \sqrt{\left(\frac{\pi}{a}\right)}\ F(z,c;\pm a) \qquad (5.36b)$$

Application of Hooke's law for the anisotropic material, eq. (3.22), to the strain intensity factors results in stress intensity factors at each crack tip (depending on which one is used in the limits).

$$\begin{aligned} K_I &= \hat{C}_{21}K_x^\varepsilon + \hat{C}_{22}K_y^\varepsilon + 2\hat{C}_{26}K_{xy}^\varepsilon \\ K_{II} &= \hat{C}_{61}K_x^\varepsilon + \hat{C}_{62}K_y^\varepsilon + 2\hat{C}_{66}K_{xy}^\varepsilon \end{aligned} \qquad (5.37)$$

5.4 Linear Variation Boundary Element Implementation

Application of the special Green's function method to the accurate fracture mechanics analysis of finite geometries with cracks has been very successful. The numerical implementation of a linear boundary element model has proven to be quite satisfactory for most problems of engineering interest. Beam bending problems (geometry aspect ratios on the order of ten or greater) are not solved as accurately without the addition of many boundary elements. A quadratic boundary data model is recommended for those aspect ratios in bending stress fields.

The geometry is to be a set of linear segments so that the integrations of the Green's functions can be performed exactly. Exact integration is generally faster than numerical integration for a level of reasonable numerical accuracy. Consider the boundary element in Fig. 5.2.

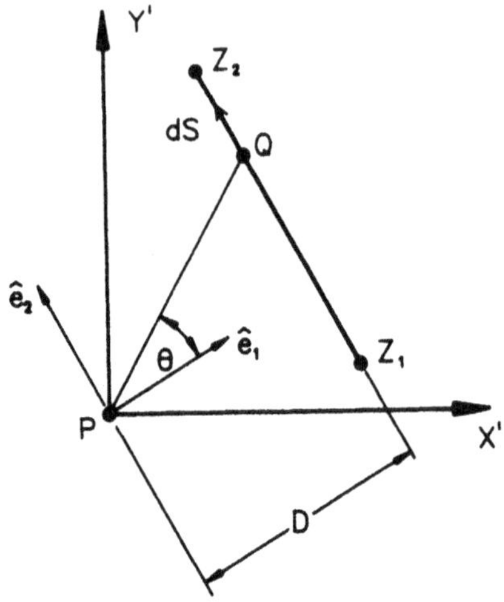

Figure 5.2. Local coordinates for piecewise linear integration

The origin of the coordinate system (x',y') is taken to be the point of application of the point load $p(x',y')$. The axes (x',y') are taken for convenience to be parallel to (x,y) centered at the crack, as shown in Fig. 5.1. However, transformation of these results to rotated coordinates is straightforward.

The complex coordinates of the source point $p(x',y')$ and of the field, or integration, point $Q(x',y')$ are given as follows

$$\begin{aligned} c_k &= x_p' + \mu_k y_p' \\ z_k &= x_Q' + \mu_k y_Q' \end{aligned} \tag{5.38}$$

where the distinct characteristic roots μ_k are given by solution of eq. 3.35.

The local coordinates in Fig. 5.2 can be used to construct the vector from $p(x',y')$ to $Q(x',y')$

$$z_k - c_k = D(A_k + B_k \tan\theta) \tag{5.39}$$

where

$$A_k = \mu_k e_1(2) + e_1(1)$$
$$B_k = \mu_k e_1(1) - e_1(2) \quad (5.40)$$

The basis vector $\hat{e}_1$ is the outward normal direction to the line segment from Q_1 to Q_2 and has components with respect to x',y' of $e_1(1),e_1(2)$.

The differential line segment for the integral equations is also determined from the Fig. 5.2 construction as

$$B_k dS = dz_k \quad (5.41)$$

The linear interpolation algorithm in eq. (3.75) can then be written as

$$f_i(z_Q) = \sum_{m=1}^{2} \{[\tfrac{1}{2} - (-1)^m \frac{z_{\bar{Q}}}{\Delta z} + (-1)^m \frac{z_Q}{\Delta z}] f_i^{\,m}\} \quad (5.42)$$

where $f_i(z_Q)$ is the interpolation vector field variable, $f_i^{\,m}$ are the end point values for the linear interpolation model and

$$z_{\bar{Q}} = (z_1+z_2)/2$$
$$\Delta z = z_2-z_1 \quad (5.43)$$

The interpolation of the boundary displacements is taken to be continuous, while the boundary tractions are allowed to be discontinuous. As discussed in Section 3.8, the tractions may have only as many as one or two unknowns per node, depending on whether or not the corresponding nodal displacements are specified. The numerical algorithm will be based on nodal displacements and segmental tractions (a value of traction vector for each end of each segment). The BIE formulation of eq. (3.78) holds for the current discussion.

Substitution of the linear algorithm of eq. (5.42) into eq. (3.78) obtains

$$C_{ij}u_j(P_N) = \sum_{M=1}^{NSEG} \{\sum_{m=1}^{2} [\tfrac{1}{2} \int_{\Delta S_m} T_{ij}(P_N,Q)dS_Q - (-1)^m \frac{z_{\bar{Q}}}{\Delta z} \int_{\Delta S_m} T_{ij}(P_N,Q)dS_Q$$

$$+ (-1)^m \frac{1}{\Delta z} \int_{\Delta S_m} z_Q T_{ij}(P_N,Q)dS_Q]\, u_j^m\} = \sum_{M=1}^{NSEG} \{\sum_{m=1}^{2} [\tfrac{1}{2} \int_{\Delta S_m} U_{ij}(P_N,Q)dS_Q$$

$$(5.44)$$

$$- (-1)^m \frac{z_{\bar{Q}}}{\Delta z} \int_{\Delta S_m} U_{ij}(P_N,Q)\, dS_Q + (-1)^m \frac{1}{\Delta z} \int_{\Delta S_m} z_Q\, U_{ij}(P_N,Q)\, dS_Q]t_j^{\,m}\}$$

where the kernels T_{ij} and U_{ij} are given in eq. (5.25) and eq. (5.21). The boundary term $C_{ij}(P_N)$ is dependent on the shape of the boundary and will be computed directly in a manner to be described further in this section.

The integrals in eq. (5.44) will be performed in closed form using complex variable integration. The integrals will be for a fixed source point node (N) and segment (M). The case where the source point is at one or the other end of the segment will be treated separately.

$$\Delta^0 T^0(P_N,Q^M;k) = \int_{\Delta S_M} \frac{dS}{z_k - c_k} = \frac{1}{B_k}\int_{z_1}^{z_2} \frac{dz_k}{z_k - c_k} \tag{5.45}$$

$$\Delta^0 T^0(P_N,Q^M;k) = \log(z_k - c_k)/B_k \Big|_{z_1}^{z_2} \tag{5.46}$$

where the symbol Δ^0 refers to those integrals in eq. (5.44) where z_Q does not occur (or occurs to the zero power). The symbol Δ^1 refers to the linear interpolation integrals, such as

$$\Delta^1 T^0(P_N, Q^M;k) = \frac{1}{B_k}\int_{z_1}^{z_2} \frac{z_k dz_k}{z_k - c_k}$$

$$\Delta^1 T^0(P_N,Q^M;k) = \frac{1}{B_k}\left(z_k \Big|_{z_1}^{z_2} + c_k \log(z_k - c_k)\Big|_{z_1}^{z_2}\right) \tag{5.47}$$

The Green's function terms in eq. (5.25) can be integrated in a single form given for the piecewise constant integral

$$\Delta^0 T^*(P_N,Q^M;k) = \frac{1}{B_k}\int_{z_1}^{z_2} \frac{\partial J(z_k,c)}{\partial z_k} dz_k$$

$$= \frac{1}{B_k} J(z_k,c) \Big|_{z_1}^{z_2} \tag{5.48}$$

and for the linear variation integral by

$$\Delta^1 T^*(P_N, Q^M;k) = \frac{1}{B_k} \int_{z_1}^{z_2} z_k \frac{\partial J(z_k,c)}{\partial z_k} dz_k \qquad (5.49)$$

$$= \frac{1}{B_k} \left\{cJ(z_k,c) + z_k - \surd(z_k^2-a^2) - I(c)\log[z_k+\surd(z_k^2-a^2)]\right\} \Big|_{z_1}^{z_2}$$

where $c = (c_k,\bar{c}_\ell,\bar{c}_k)$ as in eq. (5.25) and where $I(c)$ is given by eq. (5.17).

Following the same procedures we may integrate the displacement kernels to obtain

$$\Delta^0 U^0 (P_N,Q^M;k) = \frac{1}{B_k} [z_k \log z_k - z_k] \Big|_{z_1}^{z_2} \qquad (5.50)$$

$$\Delta^1 U^0 (P_N,Q^M;k) = \frac{1}{2B_k} [z_k^2 \log z_k - z_k^2/2] \Big|_{z_1}^{z_2} \qquad (5.51)$$

$$\Delta^0 U^* (P_N,Q^M;k) = \frac{1}{B_k} \{(z_k-c)J(z_k,c) - z_k + \surd(z_k^2-a^2)$$

$$+ I(c) \log [z_k+\surd(z_k^2-a^2)]\} \Big|_{z_1}^{z_2} \qquad (5.52)$$

$$\Delta^1 U^*(P_N,Q^M;k) = \frac{1}{B_k} \{(z_k-c)^2 J(z,c)/2 + c\, \Delta^0 U^* B_k$$

$$+ c[z_k-\surd(z_k^2-a^2) - I(c)\log (z_k+\surd(z_k^2-a^2)]/2 - z_k^2/4 \qquad (5.53)$$

$$+ [\surd(c^2-a^2)\ \surd(z_k^2-a^2)]/2 - c\surd(z_k^2-a^2)/2 + z_k\surd(z_k^2-a^2)/4$$

$$+ a^2\log[z_k + \surd(z_k^2-a^2)]/4\} \Big|_{z_1}^{z_2}$$

For later convenience let us take the following notation for these integrals and related terms in eq. (5.44) and (5.25)

$$\Delta T_{ji}^{0} = \sum_{k=1}^{2} (Q_{ik}B_k A_{jk}(\Delta^0 T^0) - \frac{Q_{ik}B_k}{2\pi(\mu_\ell-\mu_k)} \{(\mu_\ell-\mu_k)A_{jk}[\Delta^0 T^*(z_k,c_k)]$$

$$+ (\mu_\ell-\bar{\mu}_\ell)\bar{A}_{j\ell}\,[\Delta^0 T^*(z_k,\bar{c}_\ell)] + (\mu_\ell-\bar{\mu}_k)\bar{A}_{jk}\,[\Delta^0 T^*(z_k,\bar{c}_k)]\}) \;;\; \ell=3-k \tag{5.54}$$

and from eq. (5.18) and eq. (5.21)

$$\Delta U_{ji}^{0} = \sum_{k=1}^{2} (P_{ik}A_{jk}(\Delta^0 U^0) - \frac{P_{ik}}{2\pi(\mu_\ell-\mu_k)} \{+ (\mu_\ell-\mu_k)A_{jk}\,[\Delta^0 U^*(z_k,c_k)]$$

$$+ (\mu_\ell-\bar{\mu}_\ell)\bar{A}_{j\ell}[\Delta^0 U^*(z_k,\bar{c}_\ell)] + (\mu_\ell-\bar{\mu}_k)\bar{A}_{jk}\,[\Delta^0 U^*(z_k,\bar{c}_k)]\}) \quad ; \quad \ell=3-k \tag{5.55}$$

Further, let ΔT_{ji}^{1}, ΔU_{ji}^{1} be obtained by the same forms as eq. (5.54) and (5.55), except that

$$\Delta^1 T \leftarrow z_{\bar{Q}}\, \Delta^0 T/\Delta z$$
$$\Delta^1 U \leftarrow z_{\bar{Q}}\, \Delta^0 U/\Delta z \tag{5.56}$$

from the boundary interpolation in eq. (5.43). Also let ΔT_{ji}^{2} and ΔU_{ji}^{2} be obtained as in eq. (5.54) and (5.55), except that

$$\Delta^2 T \leftarrow \Delta^1 T/\Delta z$$
$$\Delta^2 U \leftarrow \Delta^1 U/\Delta z \tag{5.57}$$

Then, from the boundary interpolation, eq. (5.42), we may define

$$\Delta T_{ji} = \mathrm{Re}\,\{\Delta T_{ji}^{0} - 2(-1)^m\,[\Delta T_{ji}^{1} - \Delta T_{ji}^{2}]\} \tag{5.58}$$

and

$$\Delta U_{ji} = \text{Re}\ \{\Delta U_{ji}^{0} - 2(-1)^{m}\ [\Delta U_{ji}^{1} - \Delta U_{ji}^{2}]\} \tag{5.59}$$

resulting in a simplified form of the discretized boundary-integral equation from eq. (5.44)

$$C_{ji}(P_N)u_i(P_N) + \sum_{M=1}^{NSEG} \{\sum_{m=1}^{2} (\Delta T_{ji}u_i^m)\} = \sum_{M=1}^{NSEG} \{\sum_{m=1}^{2} (\Delta U_{ji}t_i^m)\} \tag{5.60}$$

It is now necessary to discuss the evaluation of the free term coefficient $C_{ji}(P_N)$. As first shown by Cruse (1974), eq. (5.60) may be written for the zero traction case as

$$C_{ji}(P_N)\hat{u}_i(P_N) + \sum_{M=1}^{NSEG} \{\sum_{m=1}^{2} (\Delta T_{ji}\hat{u}_i^m)\} = 0 \tag{5.61}$$

The hats on the displacements in eq. (5.61) are for the rigid body displacement fields corresponding to the zero stress case (no applied tractions). If the displacements at the nodes are taken to be unit rigid translations in the x,y directions as

$$\{\hat{u}_i^m\} = \{{\delta_{i1} \atop \delta_{i2}}\} \tag{5.62}$$

then

$$C_{ji}(P_N) = - \sum_{M=1}^{NSEG} \{\sum_{m=1}^{2} (\text{Re}\Delta T_{ji}^{0})\} \tag{5.63}$$

The evaluation of the integrals in the terms ΔT_{ji}^{0} for the source point in the segment of integration are performed in the usual Cauchy principal value sense. That is, the integral excludes a small segment of vanishing length from the end of the segment. A careful study of this process in two dimensions shows that the parameter (say ε) describing the length of the excluded segment adjacent to P_N cancels from the BIE discretization equations as the exclusion lengths from both sides of P_N are accounted for.

Special care is needed for the case of $P_N \varepsilon \Gamma$. The problem is best understood in terms of Fig. 5.3, where both crack surfaces are separately shown. P_N is restricted to exclude being placed at either crack tip, x = ±a. In Fig. 5.3, the regular surface **S** is taken to cross Γ in the manner shown. The approach is appropriate for any manner of intersection of **S** and Γ, not at the crack tips, x = ±a. The complex coordinate for P_N, denoted c_N, is shown, along with its conjugate position $\bar{c}_N$.

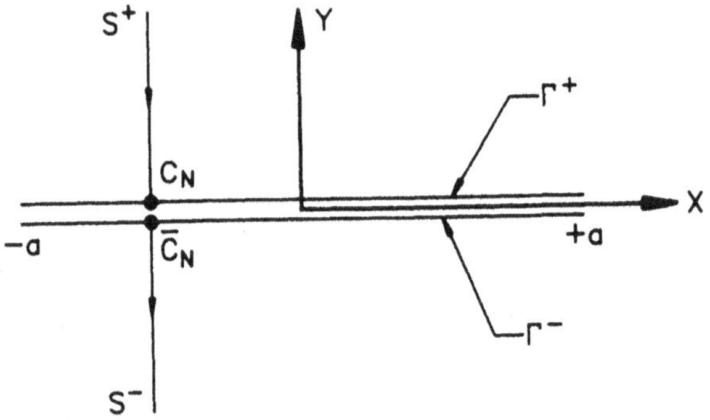

Figure 5.3. Limiting approaches to crack surface

Consider the Green's function term from eq. (5.20)

$$J(z,c) = \pi \log \left[\frac{\sqrt{(z^2-a^2)}\sqrt{(c^2-a^2)} + cz-a^2}{[z-\sqrt{(z^2-a^2)}][c+\sqrt{(c^2-a^2)}]}\right] \tag{5.64}$$

Let z = c+Δz where Δz approaches the crack surface from the opposite side from the source point c; that is

$$\sqrt{(z^2-a^2)} = -\sqrt{(c^2-a^2)} \tag{5.65}$$

for Δz→o. Expansion of eq. (5.64) for this case results in the limiting form for J(z,c) for small Δz

$$J(z,c) \to 2 \log \left(\frac{\Delta z}{\pm i\sqrt{(a^2-x^2)}}\right) \qquad x<a \tag{5.66}$$

The sign of $i = \sqrt{-1}$ is to be taken in conformance with the branch cut for J(z,c): $-\pi<\arg\leq\pi$,

$$o < \arg(z-c) < \pi \qquad J(z,c) = 2\pi \log \left(\frac{\Delta z}{i\sqrt{(a^2-x^2)}}\right)$$

$$(5.67)$$

$$-\pi < \arg(z-c) < o \qquad J(z,c) = 2\pi \log \left(-\frac{\Delta z}{i\sqrt{a^2-x^2}}\right)$$

There are two cases of concern for the principal value integrals at the crack. In the first case, the point load is on one side of the crack (say on S^+) and the field point is on the segment across from the source point (S^-). The log-term in ΔT^o_{ji} has a singularity for this condition, even though the point load is only applied to the upper crack surface. The Green's function terms in ΔT^*_{ji} from eq. (5.67) contain the cancelling singularity. For the described case, the appropriate form for the ΔT^*_{ji} is the second one in eq. (5.67)

$$\Delta T_{ji} = \sum_{k=1}^{2} \{Q_{ik}A_{jk} \log(\Delta z) + \frac{Q_{ik}}{2\pi(\mu_k-\mu_\ell)} [(\mu_\ell-\mu_k)A_{jk}$$

$$2\pi\log\left(-\frac{\Delta z}{i\sqrt{(a^2-x^2)}}\right) + \text{Regular Terms}]\} ; \qquad \ell=3-k \qquad (5.68)$$

As can be seen by inspection, the two $\log(\Delta z_k)$ terms cancel for this case.

The second case is the principal value problem when the source point and the field point are on the same side of the crack. In this case the second and third terms in ΔT^*_{ji} are singular

$$\Delta T_{ji} = \sum_{k=1}^{2} \{Q_{ik} A_{jk} \log(\Delta z)_k + \frac{Q_{ik}}{2\pi(\mu_k-\mu_\ell)} [\text{Regular term}$$

$$+ 2\pi(\mu_\ell-\bar{\mu}_\ell)\, \bar{A}_{j\ell} \log\left(\frac{\Delta z_k}{i\sqrt{(a^2-x^2)}}\right) + 2\pi(\mu_\ell-\bar{\mu}_k)\, \bar{A}_{jk} \log\left(\frac{\Delta z_k}{i\sqrt{(a^2-x)^2}}\right)]\} \qquad (5.69)$$

The terms in $\log(\Delta z_k)$ can be grouped, using the terms in eq. (3.45) for the A_{jk} quantities to obtain

$$\Delta T_{ji} = \sum_{k=1}^{2} \{Q_{ik} \frac{\mu_\ell \delta_{j2} + \delta_{j1}}{2\pi i(\mu_\ell-\mu_k)} \log(z_k-c_k) + \text{Regular Terms}\} \qquad (5.70)$$

The coefficient of the log term can be shown to properly reflect a corner singularity at Γ, for S intersecting at right angles.

Numerical implementation to compute $C_{ji}(P_N)$ for these cases is as before, but this time neglecting all log terms at the singularity and using the proper limiting forms for J(z,c), as taken from eq. (5.67), without the Δz contribution.

The Somigliana identity calculations for interior strains and the stress intensity factor are calculated in the same manner as the terms for the BIE implementation. The key terms for the strain kernels in eq. (5.34) are obtained by differentiation of the T_{ji} and U_{ji} kernels with respect to the source point coordinates. Beginning with U_{ji}, we differentiate with respect to **x** or **y** (denoted ∂x,∂y)

$$\frac{\partial U_{ji}}{\partial x,\partial y} = 2\mathrm{Re}\ \{\sum_{k=1}^{2} [-P_{ik}\ A_{jk}(\frac{1}{z_k-c_k})(1,\mu_k) + \frac{P_{ik}}{2\pi(\mu_k-\mu_\ell)}\ \{(\mu_\ell-\mu_k)$$

$$A_{jk}\ \acute{J}\ (z_k,c_k)\ (1,\mu_k) + (\mu_\ell-\bar{\mu}_\ell)\ \bar{A}_{j\ell}\acute{J}\ (z_k,\bar{c}_\ell)\ (1,\bar{\mu}_\ell) \tag{5.71}$$

$$+ (\mu_\ell-\bar{\mu}_k)\bar{A}_{jk}\ \acute{J}\ (z_k,\bar{c}_k)\ (1,\bar{\mu}_k)\}]\}$$

where $\acute{J}(z,c) = \partial J/\partial c$.

The necessary boundary integrals for the functions within eq. (5.71) can be computed as before

$$\Delta^0U^{\varepsilon *}(p,Q^M;k) = \frac{1}{B_k}\ \log\ (z_k-c)\ \Big|_{z_1}^{z_2} \tag{5.72}$$

$$\Delta^0U^{\varepsilon *}(p,Q^M;k) = \frac{1}{B_k}\ (z_k-c)\ \acute{J}-J + \frac{\partial I}{\partial c}\ \log\ [z_k+\sqrt{(z_k^2-a^2)}]\ \Big|_{z_1}^{z_2} \tag{5.73}$$

$$\Delta^1U^{\varepsilon 0}(p,Q^M;k) = \frac{1}{B_k}\ [z_k + c_k\log(z_k-c)]\ \Big|_{z_1}^{z_2} \tag{5.74}$$

$$\Delta^1 U^{\varepsilon *}(p,Q^M;k) = \frac{1}{B_k}\Big[-(z_k-c)J + (z_k-c)^2 J' + B_k \Delta^0 U^*$$

$$+ c\partial(\Delta^0 U^*)/\partial c - (z_k + \sqrt{(z^2-a^2)} \tag{5.75}$$

$$+ I(c)\log[z_k+\sqrt{(z_k^2-a^2)}])/2 - c(\partial I/\partial c)$$

$$\log\,[z_k+\sqrt{(z_k^2-a^2)}]/2 + \frac{c\sqrt{(z_k^2-a^2)}}{2\sqrt{(c^2-a^2)}} + \frac{1}{2}\ \sqrt{(z_k^2-a^2)}\Big]\ \Big|_{z_1}^{z_2}$$

The derivatives of the traction kernels for the Somigliana strain identity are denoted

$$\frac{\partial T_{ji}}{\partial x,\partial y} = 2\mathrm{Re}\ \Big\{\sum_{k=1}^{2}\ \Big[Q_{ik}\ B_k\ A_{jk}\frac{1}{(z_k-c_k)^2}\ (1,\mu_k) \tag{5.76}$$

$$+ \frac{Q_{ik}\ B_k}{2\pi(\mu_k-\mu_\ell)}\ \{(\mu_\ell-\mu_k)\ A_{jk}\ \frac{\partial J'}{\partial z_k}\ (1,\mu_k) + (\mu_\ell-\bar{\mu}_\ell)\ \bar{A}_{j\ell}\ \frac{\partial J'}{\partial z_k}\ (1,\bar{\mu}_\ell)$$

$$+ (\mu_\ell-\bar{\mu}_k)\bar{A}_{jk}\ \frac{\partial J'}{\partial z_k}\ (1,\bar{\mu}_k)\}\Big]\Big\}$$

The necessary boundary integrals for these functions are given by

$$\Delta^0 T^{\varepsilon *}(p,Q^M;k) = \frac{1}{B_k}\ \frac{1}{z_k-c_k}\ \Big|_{z_1}^{z_2} \tag{5.77}$$

$$\Delta^0 T^{\varepsilon *}(p,Q^M;k) = \frac{1}{B_k}\ J'\ \Big|_{z_1}^{z_2} = \frac{1}{\sqrt{(c^2-a^2)}\sqrt{(z_k^2-a^2)}+cz_k-a^2}\ \Big(\frac{c\sqrt{(z_k^2-a^2)}}{\sqrt{(c^2-a^2)}}\Big) \tag{5.78}$$

$$+ z_k\ \Big|_{z_1}^{z_2}$$

$$\Delta^1 T^{\varepsilon 0}(p,Q^M;k) = \frac{1}{B_k}\ \Big[\log\ (z_k-c_z) - \frac{c}{z_k-c}\Big]\ \Big|_{z_1}^{z_2} \tag{5.79}$$

$$\Delta^{1}T^{\varepsilon *}(p,Q^{M};k) = \frac{1}{B_k}\{J + CJ' - \frac{\partial I}{\partial c}\log[z_k+\sqrt{(z_k^2-a^2)}]\}\Big|_{z_1}^{z_2} \qquad (5.80)$$

The directional derivatives in eq. (5.71) and (5.76) are denoted therein in parenthetical terms. If we denote the direction of the derivative as $x_r(r=1,2)$ then we can define material terms

$$\{R_{rk}\} = \{{1 \atop \mu_k}\} \qquad (5.81)$$

$$\{S_{rk}\} = \{\frac{1}{\mu_k}\} \qquad (5.82)$$

The strain kernels are then written, following the notation, of eq. (5.54) as

$$\Delta T^{\varepsilon 0}_{jir} = \sum_{k=1}^{2}[B_k A_{jk}(\Delta^{0}T^{\varepsilon 0})(Q_{ik}R_{rk}+Q_{rk}R_{ik})$$

$$-\frac{B_k}{2\pi(\mu_\ell-\mu_k)}\{(\mu_\ell-\mu_k)A_{jk}(\Delta^{0}T^{\varepsilon *})(Q_{ik}R_{rk}+Q_{rk}R_{ik}) \qquad (5.83)$$

$$+(\mu_\ell-\bar{\mu}_\ell)\bar{A}_{j\ell}(\Delta^{0}T^{\varepsilon *})(Q_{ik}S_{r\ell}+Q_{rk}S_{i\ell})+(\mu_\ell-\bar{\mu}_k)\bar{A}_{jk}(\Delta^{0}T^{\varepsilon *})$$

$$(Q_{ik}S_{rk}+Q_{rk}S_{ik})\}] \qquad ;\ell=3-k$$

and as

$$\Delta U^{\varepsilon 0}_{jir} = \sum_{k=1}^{2}[A_{jk}(\Delta^{0}U^{\varepsilon 0})(P_{ik}R_{rk}+P_{rk}R_{ik})$$

$$-\frac{1}{2\pi(\mu_\ell-\mu_k)}\{(\mu_\ell-\mu_k)A_{jk}(\Delta^{0}U^{\varepsilon *})(P_{ik}R_{rk}+P_{rk}R_{ik}) \qquad (5.84)$$

$$+(\mu_\ell-\bar{\mu}_\ell)\bar{A}_{j\ell}(\Delta^{0}U^{\varepsilon *})(P_{ik}S_{r\ell}+P_{rk}S_{i\ell})+(\mu_\ell-\bar{\mu}_k)\bar{A}_{jk}(\Delta^{0}U^{\varepsilon *})$$

$$(P_{ik}S_{rk}+P_{rk}S_{ik})\}] \qquad ;\ell=3-k$$

The additional terms to complete the linear variation algorithm for internal strains are given by the same substitutions as in eq. (5.56), (5.57) using the terms in (5.74), (5.75) and (5.79), (5.80). The total strain kernels are then assembled as in eq. (5.58), (5.59), such that

$$\varepsilon_{ji}(p) = \sum_{M=1}^{NSEG} \{\sum_{m=1}^{2} (\Delta\, T^{\varepsilon}_{jir}\, u^{m}_{r})\} - \sum_{M=1}^{NSEG} \{\sum_{m=1}^{2} (\Delta\, U^{\varepsilon}_{jir}\, t^{m}_{r})\} \tag{5.85}$$

The necessary integrals for evaluating the strain terms from eq. (5.36b) are derived from the limiting operator on eqs. (5.73), (5.75), (5.78), and (5.80). It may be shown that the limiting process for the x = +a crack tip results in the following terms

$$R^{0}(k) \triangleq \lim_{R\to 0} (\sqrt{2\pi R}\, \Delta\, {}^{0}U^{\varepsilon *}) = \frac{1}{B_k} \sqrt{(\tfrac{\pi}{2})}\, (\sqrt{[(z/a)^2-1]} + \log\{(z/a)+\sqrt{[(z/a)^2-1]}\})\Big|_{z_1}^{z_2} \tag{5.86}$$

$$R^{1}(k) \triangleq \lim_{R\to 0} (\sqrt{2\pi R}\, \Delta\, {}^{1}U^{\varepsilon *}) = \frac{1}{B_k} \sqrt{(\tfrac{\pi}{a})}\, \{(\tfrac{z}{2a} + 1)\sqrt{(z^2-a^2)} + \tfrac{a}{2} \log[z + \sqrt{(z^2-a^2)}]\}\Big|_{z_1}^{z_2} \tag{5.87}$$

$$L^{0}(k) \triangleq \lim_{R\to 0} (\sqrt{2\pi R}\, \Delta\, {}^{0}T^{\varepsilon *}) = -\frac{1}{B_k} \sqrt{(\tfrac{\pi}{2})}\, (\tfrac{1}{a} \tfrac{\sqrt{(z+a)}}{\sqrt{(z-a)}})\Big|_{z_1}^{z_2} \tag{5.88}$$

$$L^{1}(k) \triangleq \lim_{R\to 0} (\sqrt{2\pi R}\, \Delta\, {}^{1}T^{\varepsilon *}) = \frac{1}{B_k} \sqrt{(\tfrac{\pi}{2})}\, \{\log\, [z + \sqrt{(z^2-a^2)}] - \frac{\sqrt{(z+a)}}{\sqrt{(z-a)}}\}\Big|_{z_1}^{z_2} \tag{5.89}$$

Substitution of the results in eqs. (5.86)-(5.89) into the piecewise constant terms in eqs. (5.83) and (5.84), and into their linear equivalents, provides the strain intensity factor integrals for eq. (5.38) in order to obtain K_I and K_{II} at the crack tip (x = +a). Similar results can be obtained for the negative crack tip stress intensity factors.

5.5 Applications

The computer implementation of the integrals in Section 5.4 has been accomplished and is available for Personal Computer or mainframe analysis. The code is referred to as BIE/CRX for crack analysis. While the current code does not include quadratic data variation and curved surface elements, it has been found to give exceptionally good accuracy for many engineering applications.

The first benchmark problems are shown in Fig. 5.4 for the center-crack and edge crack configurations. These models are for a/w=0.5. Various BEM models are shown for the symmetric part of each geometry. Table 5.1 compares the BIE/CRX results for these models to the collocation results of Tada (1973). Tada's results are cited as being within 0.5% of the estimated true value. The BIE/CRX models were selected to be crude, but the comparison to Tada is quite satisfactory. The edge crack geometry involves significant rotation effects in the displacements, which are not as well modeled for the piecewise linear data model used. It can be argued that the use of a quadratic integral term added to the linear terms in eq. (3.78) would bring the edge crack accuracy to that of the center crack results. The same effect can be achieved with many more boundary elements, but convergence is slow.

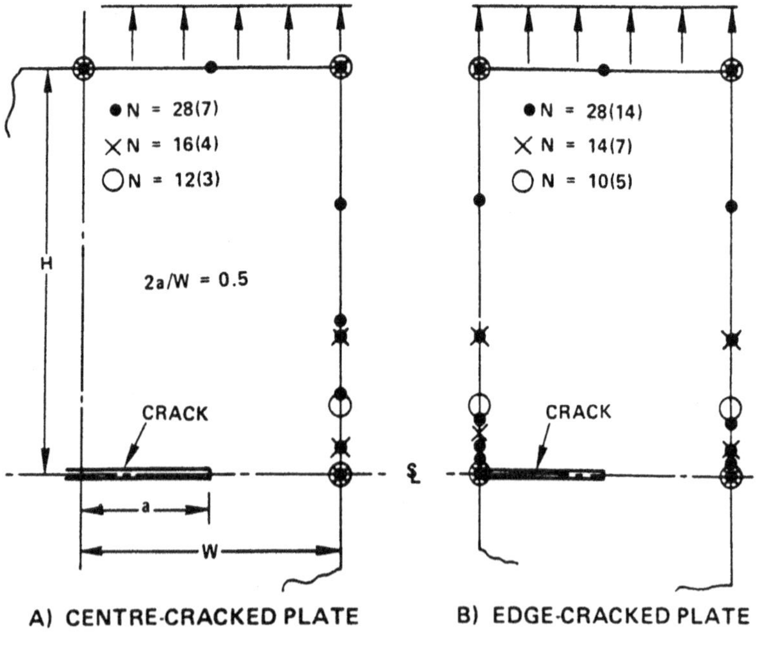

Figure 5.4. Benchmark BIE/CRX models

Figure 4.3 shows the problem of a tension plate containing an angle crack. Three BIE/CRX models were used with 16, 24, and 40 nodes on S without the multiregion mesh on the interior. Table 5.2 demonstrates the BIE/CRX convergence results for this problem.

Table 5.1. Benchmark BIE/CRX modeling results

	N	$K_1/\sigma\sqrt{\pi a}$	Time (sec)	% Diff.
Center-crack	12(3)	1.169	1.5	1.6
	16(4)	1.180	2.4	0.6
	28(7)	1.196	6.4	-0.7
Edge-crack	10(5)	2.703	1.7	4.6
	14(7)	2.697	3.2	4.9
	28(14)	2.791	11.8	1.3

Table 5.2. Mixed-mode stress intensity factor data from BIE/CRX

No. Nodes	$K_I/\sigma\sqrt{\pi a}$	$K_{II}/\sigma\sqrt{\pi a}$	% Deviation K_I	% Deviation K_{II}
16	0.671	0.562	+5.7	+4.8
24	0.710	0.577	+0.2	+2.3
40	0.712	0.590	-	-

Figure 5.5 is for a tension plate with symmetry along both the x and y-axes, although the y-axis symmetry is simulated with rollers. The BEM model contains 26 nodes and treats the hole as a series of straight-line segments. The BIE/CRX model for zero crack length was determined from the strain in the smallest boundary segment near the x-axis to be $K_T = 3.29$. Table 5.3 summarizes the obtained stress intensity factor results for various crack sizes. Each result is obtained for the same mesh by changing only the crack length parameter in the BIE/CRX code.

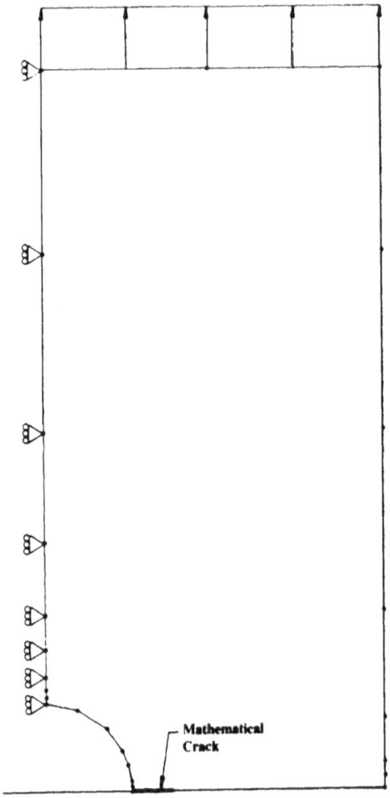

Figure 5.5. BIE/CRX model for edge cracks at circular hole

Table 5.3. Stress intensity factors for crack at hole

a/R	$K_I/\sigma\sqrt{\pi a}$
0.05	3.30
0.25	2.49
0.50	2.06
1.00	1.80
1.50	1.83

The fracture mechanics model considered is for symmetric edge cracks at the (x = ±R, y = 0) locations. The BIE/CRX result can be obtained by letting a>R and centering the crack at (x = 0, y = 0). It is also possible to place the center of the crack at (x = R, y = 0) and get essentially the same numerical results.

More recently, the BIE/CRX code was applied to the fracture mechanics analysis of single crystals by Chan and Cruse (1986). The cracks grow in directions set by the slip systems in the single crystals, and these are generally inclined to the specimen axes. Figures 5.6-5.7 show two BEM models, both using 43 nodes, for different angled crack conditions. The crack length (2a) is positioned in these models of a compact tension specimen such that the negative crack tip is just external to the modeled region (in the open notch), while the remainder of the crack length extends into the specimen, as shown. This modeling approach was validated for the $\beta=0^0$ case and for various crack lengths. Results for K_I and K_{II} were obtained from this model and used by Chan to generate crack growth rate models.

It is seen that the BIE/CRX models in Figs. 5.5-5.7 eliminate the need to model the crack. Further, the boundary element models are simple and easily constructed. While some modeling change is made between the two maps in Figs. 5.6 and 5.7, these are of a minor nature.

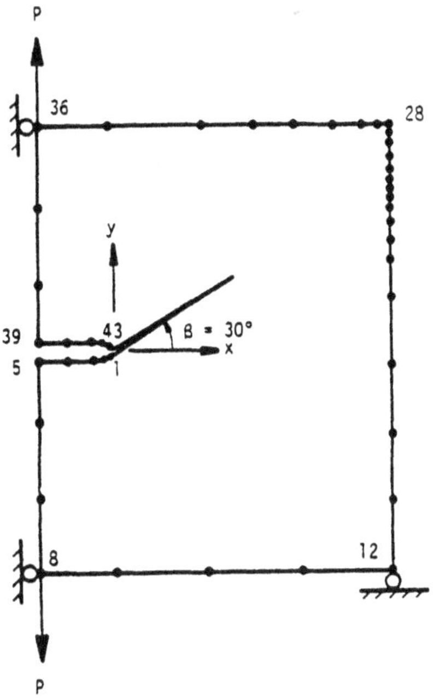

Figure 5.6. Compact tension model with 30° angled crack

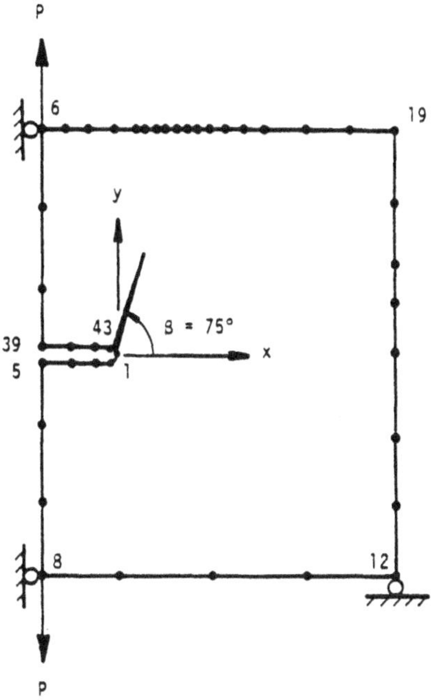

Figure 5.7. Compact tension model with 75° angled crack

References

(1971) A.H. England, Complex Variable Methods in Elasticity, Wiley Interscience, London.

(1971) M.D. Greenberg, Application of Green's Functions in Science and Engineering, Prentice-Hall, Englewood Cliffs, N.J.

(1973) M.D. Snyder, Crack Tip Stress Intensity Factors in Finite Anisotropic Plates, Ph.D. Dissertation, Carnegie-Mellon University, Pittsburgh, PA.

(1973) H. Tada, The Stress Analysis of Cracks Handbook, Del Research Corporation, Hellertown, PA.

(1975) M.D. Snyder and T.A. Cruse, Boundary-integral Equation Analysis of Cracked Anisotropic Plates, A Contour Integral Computation of Mixed-Mode Stress Intensity Factors, International Journal of Fracture, 11, 315-328.

(1978) T.A. Cruse, Two Dimensional BIE Fracture Mechanics Analysis, Applied Mathematical Modeling, 2, 287-293.

(1986) K.S. Chan and T.A. Cruse, Stress Intensity Factors for Anisotropic Compact-Tension Specimens with Inclined Cracks, Engineering Fracture Mechanics, 23, 5, 863-874.

6.0 Elastoplastic Fracture Mechanics Analysis

6.1 Introduction

The developments so far have been limited to elastic material behavior. However, it is well understood that, for many problems, the plastic response of the near-crack tip material results in local damage that is not described accurately by linear elastic fracture mechanics analysis. While the stresses are limited to finite values due to plastic effects, the local strain intensity will generally be greater than the elastic case. While non-continuum slip or void formation damage may occur, the present discussion is limited to classical continuum behavior.

BIE analysis is most properly seen as an elastic methodology. Inelastic response will be seen to require volume (three-dimensional) or two-dimensional) discretization in addition to boundary modeling. The substantial advantage of the BEM over the FEM for fracture mechanics analysis may be much less for the elastoplastic case. However, use of the special Green's function formulation of Chapter 5 results in some significant insight for the two-dimensional elastoplastic fracture mechanics problem. Therefore, that approach is now to be developed.

The elastoplastic formulation for the BIE method was first developed by Swedlow and Cruse (1971) for three dimensions. Riccardella (1973) made a first, and partially successful, numerical implementation in two dimensions, as did Mendelson (1973). The two-dimensional elastoplastic formulation was corrected by Mukherjee (1977). Bui (1978) added further, critical insight into the elastoplastic BIE formulation for the internal strain evaluations. The books by Mukherjee (1982) and Telles (1983) are major compendia of elastoplastic BIE results.

The elastoplastic BIE formulation for the fracture mechanics problem was first given by Morjaria and Mukherjee (1981) for the case of a plate with a narrow ellipse; a special Green's function was used. Cruse and Polch (1986) reported on the successful implementation of the elastoplastic BIE formulation for the special Green's function of an ideal crack, as described in Chapter 5. It is this formulation which is detailed in the following sections.

6.2 Fundamental Elastoplastic Relations

The classical theory of incremental elastoplasticity for work hardening materials will be used. Isotropic hardening is implied, but other hardening rules are easily incorporated in the final computer program.

We begin with the notion of a loading function $F(\sigma_{ij},\varepsilon^P_{ij},\kappa)$ which satisfies Prager's consistency relation, requiring that the loading from one plasticity state leads smoothly to another plasticity state. The plastic strains are denoted ε^P_{ij}, and the yield surface hardening parameter is given by κ. The selected form for the loading function is given by

$$F(\sigma_{ij},\varepsilon^P_{ij},\kappa) = \phi(S_{ij}) - \kappa(\varepsilon^P_{ij},W_p) = 0 \tag{6.1}$$

where W_p, the plastic work, is given by

$$W_p = \int_o^{\varepsilon^P_{ij}} S_{ij} d\varepsilon^P_{ij} \tag{6.2}$$

and where S_{ij} are the components of the deviatoric stress. The form of the yield surface function $\phi(S_{ij})$ precludes a mean stress dependence on yield.

For continued yielding, the incremental or slow rate change $\delta F(\sigma_{ij},\varepsilon^P_{ij},\kappa)$ must be zero, such that

$$\frac{\partial F}{\partial \sigma_{ij}} \delta\sigma_{ij} + \frac{\partial F}{\partial \varepsilon^P_{ij}} \delta\varepsilon^P_{ij} + \frac{\partial F}{\partial \kappa} \delta\kappa \equiv 0 \tag{6.3}$$

Combining these first relations we obtain

$$\frac{\partial \phi}{\partial S_{ij}} \delta S_{ij} - \frac{\partial \kappa}{\partial \varepsilon^P_{ij}} \delta\varepsilon^P_{ij} - \frac{\partial \kappa}{\partial W_p} \frac{\partial W_p}{\partial \varepsilon^P_{ij}} \delta\varepsilon^P_{ij} \equiv 0 \tag{6.4}$$

Drucker's hypothesis states that no useful work can be done (over and above the amount of elastic work) through the application of small additional external stresses and then their removal. From his hypothesis we know that the increment in plastic strain must be normal to all smooth parts of the yield surface $\phi(S_{ij})$, or

$$\delta\varepsilon^P_{ij} = \lambda \frac{\partial \phi}{\partial S_{ij}} \tag{6.5}$$

Substituting eq. (6.5) into eq. (6.4) we obtain for the magnitude λ

$$\lambda = \frac{(\partial\phi/\partial S_{ij})\ \delta S_{ij}}{(\partial\kappa/\partial\varepsilon^{P}_{\kappa\ell} + S_{\kappa\ell}\partial\kappa/\partial W_{p})(\partial\phi/\partial S_{\kappa\ell})} \tag{6.6}$$

such that

$$\delta\varepsilon^{P}_{ij} = \frac{(\partial\phi/\partial S_{mn})\delta S_{mn}(\partial\phi/\partial S_{ij})}{(\partial\kappa/\partial\varepsilon^{P}_{\kappa\ell} + S_{\kappa\ell}\ \partial\kappa/2W_{p})(\partial\phi/\partial S_{\kappa\ell})} \tag{6.7}$$

Equation (6.7) demonstrates the well-known quasilinear relationship between increments of deviatoric stress and plastic strain. Reduction of the nonlinear elastoplastic problem to a quasilinear form permits implementation of these theories into incremental analysis computer codes.

The elastic strain increment, of course, is still governed by Hooke's law

$$2\mu(\delta\varepsilon_{ij}-\delta\varepsilon^{P}_{ij}) = \delta\sigma_{ij} + \frac{\nu}{1+\nu}\ \delta\sigma_{mm}\ \delta_{ij} \tag{6.8}$$

Potential confusion is to be averted by noting the difference between incremental variables ($\delta\varepsilon,\delta\sigma$) and the Kronecker delta, δ_{ij}, in eq. (6.8). Changing stresses into deviatoric stresses, eq. (6.8) may be rewritten to obtain

$$\delta S_{ij} = 2\mu(\delta\varepsilon_{ij}-\delta\varepsilon_{mm}\delta_{ij}/3) - 2\mu\delta\varepsilon^{P}_{ij} \tag{6.9}$$

Multiplying both sides of eq. (6.9) by $(\partial\phi/\partial S_{ij})/2\mu$ and combining with eq. (6.6) we find

$$\lambda = \frac{(\partial\phi/\partial S_{mn})(\delta_{mk}\delta\varepsilon_{n\ell} - \delta_{mn}\delta_{k\ell}/3)\ \delta\varepsilon_{k\ell}}{[(\partial\phi/\partial S_{ij})(\partial\phi/\partial S_{ij}) + (\partial\kappa/\partial\varepsilon^{P}_{ij}+\partial\kappa/\partial W_{p}S_{ij})(\partial\phi/\partial S_{ij})]/2\mu} \tag{6.10}$$

or

$$\lambda = B_{ij}(S,\varepsilon^{P},\kappa)\ \delta\varepsilon_{ij} \tag{6.11}$$

The results so far are for the three-dimensional problem. Based on eq. (6.9) and (6.11), the increment in the total stress is given by

$$\delta\sigma_{ij}/2\mu = \delta\varepsilon_{ij} + \frac{\nu}{1-2\nu}\ \delta\varepsilon_{mm}\delta_{ij}-(\partial\phi/\partial S_{ij})B_{mn}\delta\varepsilon_{mn} \tag{6.12}$$

For plane strain problems (letting x_3 denote the thickness direction), eq. (6.12) continues to be valid, using a range of 2 on the indices. The plane stress result must be derived separately to obtain

$$\delta\sigma_{ij}/2\mu = \delta\varepsilon_{ij} + \frac{\nu}{1-\nu}\delta\varepsilon_{\alpha\alpha}\delta_{ij} - \left(\frac{\nu}{1-\nu}\delta_{\alpha\beta}\delta_{ij}+\delta_{\alpha i}\delta_{\beta j}\right)\lambda\left(\partial\phi/\partial S_{\alpha\beta}\right) \tag{6.13}$$

where α,β have the range of 2.

In the case of isotropic work hardening $K=K(W_p)$; taking the von Mises ellipsoid for the yield surface, $\phi=S_{ij}S_{ij}/2=J_2$, the second stress invariant, we obtain from eq. (6.6)

$$\lambda = \frac{S_{ij}\delta S_{ij}}{(\partial K/\partial W_p)S_{k\ell}S_{k\ell}} = \delta J_2/(2\kappa' J_2) \tag{6.14}$$

For the case of uniaxial loading, $J_2=\sigma_0{}^2/3$, the derivative of the work hardening parameter can be computed as

$$\kappa' = \frac{2}{3}\frac{d\sigma_0}{d\varepsilon_0^p} \triangleq 2H/3 \tag{6.15}$$

where σ_0,ε_0^p are the uniaxial stress and the plastic strain. The slope of the stress plastic strain curve is taken to be H. The associated von Mises flow rule then is given by the relation

$$\lambda = \frac{1}{2J_2}\left(\frac{S_{k\ell}}{1+H/3\mu}\right)\delta\varepsilon_{k\ell} \tag{6.16}$$

and eq. (6.12) becomes

$$\delta\sigma_{ij}/2\mu = \delta\varepsilon_{ij} + \frac{\nu}{1-2\nu}\delta\varepsilon_{mm}\delta_{ij} - \frac{S_{ij}S_{k\ell}}{2J_2(1+H/3\mu)}\delta\varepsilon_{k\ell} \tag{6.17}$$

6.3 The Somigliana Identities in Three-Dimensional Elastoplasticity

Elastoplasticity is, of course, nonlinear behavior of the material. The Navier equations of equilibrium for elastoplasticity may be obtained from eq. (6.17) by differentiation and application of the strain-displacement relationship to obtain

$$\delta u_{i,jj} + \delta u_{j,ji}/(1-2\nu) - \left[S_{ij}S_{k\ell}\delta u_{k,\ell}/J_2(1+H/3\mu)\right]_{,j} = 0 \tag{6.18}$$

The differential equation of equilibrium is linear in the increments of the unknown displacements, but clearly inhomogeneous and history dependent due to the bracketed term. A fundamental solution to eq. (6.18) does not exist. Thus, the formulation of Somigliana identities for the elastoplastic problem requires a special approach. Further, as we shall see, the formulation requires some compromises.

The application of Betti's theorem to the linear elastic problem was given in Section 3.3. It was seen that reciprocity in a form of strain energy type of variable was achieved due to the Hooke's law relation between the stress and strain variables in eq. (3.56). In order to achieve a reciprocity relationship in the elastoplastic problem, we will use the elastic fundamental solution and the elastic strain increment for the desired solution. Then it is easily proven that

$$\int_V \sigma^*_{ij}(\delta\varepsilon_{ij}-\delta\varepsilon^p_{ij})dV = \int_V \varepsilon^*_{ij}\delta\sigma_{ij}\, dV \tag{6.19}$$

The parenthetical term in eq. (6.19) is the increment in elastic strain, while the starred terms refer to the usual elastic fundamental solutions. The increments of solution for the unknown problem are used rather than total values because it is not possible in many problems to determine the total plastic strain distribution over the volume. An incremental approach which accumulates the prior history terms, as in the finite element method, achieves solutions to all reasonable problems. More details on the incremental solution will be presented in a later section.

The divergence theorem may be used for transforming the first and third terms in eq. (6.19) to surface integrals. Following the approach in Section 3.3 and taking the body force to be zero, we introduce the displacement gradients as follows

$$\int_V \sigma^*_{ij}\delta u_{i,j}dV - \int_V \sigma^*_{ij}\delta\varepsilon^p_{ij}dV = \int_V u^*_{ij}\delta\sigma_{ij}dV \tag{6.20}$$

The gradient terms, in the absence of body forces, may then be integrated to the surface such that for any interior point

$$\int_S t^*_i\delta u_i dS + \delta u_i e_i - \int_V \sigma_{ij}{}^*\delta\varepsilon^p_{ij}dV = \int_S u^*_i\delta t_i dS \tag{6.21}$$

The Somigliana displacement identity is then given by rearrangement of eq. (6.21), and the use of the fundamental solutions from eqs. (3.12) and (3.16) to obtain

$$\delta u_i(p) = -\int_S T_{ij}(p,Q)\delta u_j(Q)dS + \int_S U_{ij}(p,Q)\delta t_j(Q)dS + \int_V \Sigma_{ijk}(p,q)\ \delta\varepsilon^p_{jk(q)}\ dV \qquad (6.22)$$

The volume kernel in eq. (6.22) is the stress kernel in three dimensions given by

$$\Sigma_{ijk}(p,q) = -\ [(1-2\nu)(\delta_{ij}r_{,k}+\delta_{ik}r_{,j}-\delta_{jk}r_{,i}) + 3r_{,i}r_{,j}r_{,k}]/[8\pi(1-\nu)r^2] \qquad (6.23)$$

It is seen from eq. (6.22) that the boundary-integral equation for the elastoplastic problem may be obtained as in the elastic case. The Somigliana identity may be integrated in the principal value sense of the surface $\langle S\rangle$ and the small area S_ε surrounding $P(\underset{\sim}{x})$, the boundary point. Then eq. (6.22) becomes

$$\begin{aligned}\delta u_i(P) = &-\int_{\langle S\rangle} T_{ij}(P,Q)\delta u_j(Q)dS + \int_{\langle S\rangle} U_{ij}(P,Q)\delta t_j(Q)dS \\ &-\int_{S_\varepsilon} T_{ij}(P,Q)[\delta u_j(Q)-\delta u_j(P)]dS \\ &-\delta u_j(P)\lim_{p\to P}\int_{S_\varepsilon} T_{ij}(p,Q)dS \\ &+\int_V \Sigma_{ijk}(P,q)\ \delta\varepsilon^P_{jk}(q)dV\end{aligned} \qquad (6.24)$$

No regularization is needed for the volume term, and thus we obtain the BIE as in the elastic case

$$C_{ij}\delta u_j(P) + \int_{\langle S\rangle} T_{ij}(P,Q)\ \delta u_j(Q)dS = \int_{\langle S\rangle} U_{ij}(P,Q)\delta t_j(Q)dS + \int_V \Sigma_{ijk}(P,q)\delta\varepsilon^P_{jk}(q)dV \qquad (6.25)$$

The BIE for the elastoplastic problem is not sufficient for numerical evaluation unless the elastoplastic strains are specified. In general, these strains result from the application of the loading and must themselves be treated as unknown.

The usual approach for the elastoplastic problem is to formulate the Somigliana identity for the interior strains. The strain identity, together with the BIE, provides sufficient information for numerical analysis, as will be described later in this chapter.

In order to compute the interior strains it is necessary to differentiate eq. (6.22) at the point of load application. The only problem in that operation is the singular order of the stress kernel, from eq. (6.23). Bui (1978) was the first to properly recognize the special requirements of this differentiation. Taking the derivative of eq. (6.22) with respect to the coordinates of $x_h(p)$ we obtain

$$\delta u_{i,h} = \int_S T_{ij,h}(p,Q)\delta u_j(Q)dS - \int_S U_{ij,h}(p,Q)\delta t_j(Q)dS$$

$$\frac{\partial}{\partial x_h|_p} \int_V \Sigma_{ijk}(p,q)\delta\varepsilon^p_{jk}(q)dV \qquad (6.26)$$

The signs in the first two integrals change due to $\partial/\partial x|_p = -\partial/\partial x|_q$. The derivative of the kernel $\Sigma_{ijk}(p,q)$ in the third integral leads to a $1/r^3$ singularity, which requires special consideration.

Following Bui (1978), the last integral is rewritten in a principal value sense by deleting a ball of radius ε centered at $p(\underset{\sim}{x})$. Noting that the plastic strains inside the plastic domain are continuous, Bui obtains the result

$$\delta u_{i,h} = \int_S T_{ij,h}(p,Q)\delta u_j(Q)dS - \int_S U_{ij,h}(p,Q)\delta t_j(Q)dS$$

$$- \int_{\langle V\rangle} \Sigma_{ikj,h}(p,q)\delta\varepsilon^P_{kj}(q)dV - \int_{S_\varepsilon} \Sigma_{ikj}(p,Q)n_h\delta\varepsilon^P_{kj}(Q)dS \qquad (6.27)$$

He shows that the third integral exists as a principal value. Further, the fourth integral is due to the convection of the ball of radius ε, as $p(\underset{\sim}{x})$ is moved in differentiation. This convected integral contribution can be evaluated on the surface S_ε to obtain

$$- \int_{S_\varepsilon} \Sigma_{ikj}(p,Q)n_h\,\delta\varepsilon^P_{kj}(Q)dS = \frac{8-10\nu}{15(1-\nu)}\,\delta\varepsilon^P_{ih}(p) \qquad (6.28)$$

Equation (6.27) with the result in eq. (6.28) defines the distribution of interior displacement gradient or strain increments. The incremental equations of elastoplasticity in Section 6.2, together with the boundary integral equation (6.25), form a sufficient set of equations for an iterative-incremental solution of the field problem. The details of the solution algorithm will be given in Section 6.5.

6.4 The Somigliana Identities in Two-Dimensional Elastoplasticity

We shall begin by considering the reduction of the Betti reciprocal strain energy in eq. (6.19) to two-dimensional form. As first properly reported by Mukherjee (1977), the reduction to the plane strain form requires some attention to the three-dimensional equation.

Take x_3 to be the transverse direction for a body in plane strain such that $\varepsilon_{33}=\varepsilon_{13}=\varepsilon_{23}=0$. The same restriction holds for the strain increments as well. However, this restriction does not hold for the plastic normal strain increment

$$-\delta\varepsilon_{33}^{P} = \delta\varepsilon_{11}^{P} + \delta\varepsilon_{22}^{P} \tag{6.29}$$

which is derived from the constancy of volume hypothesis for plastic strains. Using eq. (6.29) we can rewrite eq. (6.19), expanding the terms into in-plane ($\alpha,\beta=1,2$) and out-of-plane terms

$$\begin{aligned}\int_A \overset{*}{\sigma}_{\alpha\beta}\,(\delta\varepsilon_{\alpha\beta}-\delta\varepsilon_{\alpha\beta}^{P})dA + \int_A \overset{*}{\sigma}_{33}\;\delta\varepsilon_{33}dA - \int_A \overset{*}{\sigma}_{33}\delta\varepsilon_{33}^{P}dA \\ = \int_A \overset{*}{\varepsilon}_{\alpha\beta}\delta\sigma_{\alpha\beta}dA + \int_A \overset{*}{\varepsilon}_{33}\;\delta\sigma_{33}dA\end{aligned} \tag{6.30}$$

The integrals in eq. (6.30) involving $\delta\varepsilon_{33}$ and $\overset{*}{\varepsilon}_{33}$ are identically zero for plane strain. The fundamental solution stress term $\overset{*}{\sigma}_{33}$ is replaced by its plane strain equivalent, while eq. (6.29) is used for plastic strain term to obtain

$$\int_A \overset{*}{\sigma}_{\alpha\beta}(\delta\varepsilon_{\alpha\beta}-\delta\varepsilon_{\alpha\beta}^{P})dA + \nu\int_A \overset{*}{\sigma}_{\alpha\alpha}\;\delta\varepsilon_{\beta\beta}^{P}dA = \int_A \overset{*}{\varepsilon}_{\alpha\beta}\delta\sigma_{\alpha\beta}dA \tag{6.31}$$

Using Hooke's law, it is shown that the two plastic strain terms may be written as

$$\sigma^*_{\alpha\beta}\delta\varepsilon^P_{\alpha\beta} - \nu\sigma^*_{\alpha\alpha}\delta\varepsilon^P_{\varepsilon\varepsilon} = 2\mu u^*_{\alpha,\beta}\delta\varepsilon^P_{\alpha\beta} \tag{6.32}$$

Then using the divergence theorem, eq. (6.31) reduces to the Somigliana identity for the interior displacement given for plane strain by

$$\int_S t^*_\alpha \delta u_\alpha dS + \delta u_\alpha e_\alpha - 2\mu \int_A u^*_{\alpha,\beta}\, \delta\varepsilon^P_{\alpha\beta} dA = \int_S u^*_\alpha \delta t_\alpha dS \tag{6.33}$$

For plane stress, the reduction to the displacement Somigliana identity is more straightforward. The simplification is seen in eq. (6.30) where the normal stress terms σ^*_{33} and $\delta\sigma_{33}$ are zero. Writing the fundamental stresses in terms of the displacement gradients we obtain the plane stress interior Somigliana identity

$$\int_S t^*_\alpha \delta u_\alpha dS + \delta u_\alpha e_\alpha - 2\mu \int_A \left(u^*_{\alpha,\beta} + \frac{\nu}{1-\nu} u^*_{\gamma,\gamma}\delta_{\alpha\beta}\right)\delta\varepsilon^P_{\alpha\beta} dA = \int_S u^*_\alpha \delta t_\alpha dS \tag{6.34}$$

The two-dimensional Somigliana identities can be written in the form of eq. (6.22) as

$$\begin{aligned}\delta u_\alpha(p) = &- \int_S T_{\alpha\beta}(p,Q)\delta u_\beta(Q)dS + \int_S U_{\alpha\beta}(p,Q)\delta t_\beta(Q)dS \\ &+ \int_A \Sigma_{\alpha\beta\gamma}(p,q)\delta\varepsilon^P_{\beta\gamma}(q)dA\end{aligned} \tag{6.35}$$

where, for plane strain

$$\Sigma_{\alpha\beta\gamma}(p,q) = 2\mu U_{\alpha\beta,\gamma}(p,q) \tag{6.36}$$

and for plane stress

$$\Sigma_{\alpha\beta\gamma}(p,q) = 2\mu\left[U_{\alpha\beta,\gamma} + \frac{\nu}{1-\nu} U_{\alpha\delta,\delta}\delta_{\beta\gamma}\right] \tag{6.37}$$

The two-dimensional BIE formulation is, of course, of the same form as eq. (6.25).

The interior strain identity in two dimensions may be derived in the same fashion as for three dimensions, although the convected integral gives different results. These results replace eq. (6.28) in three dimensions and are given

$$- \int_{S_\varepsilon} \Sigma_{\alpha\beta\gamma} \, n_\delta \delta\varepsilon^p_{\beta\gamma} = E_{\alpha\beta\gamma\delta} \delta\varepsilon^p_{3\gamma} \tag{6.38}$$

The planer terms in eq. (6.38) are not as simple as the three-dimensional terms in eq. (6.28). The form of this solution is herein deferred to the formulation of eq. (6.55).

6.5 The Somigliana Identities in Two-Dimensional Elastoplastic Fracture Mechanics

The elastic Green's function for two-dimensional fracture mechanics analysis was derived in Chapter 5. The fundamental solution was derived for the generally anisotropic planer problem and contained two kinds of terms. The first is the usual point load in an infinite plane, while the second is the set of terms required to create a traction free crack surface.

The Somigliana identity for the elastoplastic Green's function formulation is obtained in the same steps as for eq. (6.35). That the steps are the same is due to the fact that the point-load solution is mathematically identical for the two problems (only the elastic constants differ for isotropy versus anisotropy). The fundamental stress kernel in the domain integral in eq. (6.35) differs for the anisotropic Green's function formulation. Assuming the use of plane strain, nearly isotropic material properties in eq. (3.32), and taking the kernel for plane strain elastoplasticity from eq. (6.36), the Green's function kernel is given by

$$\Sigma_{jim} = 2G \, \mathrm{Re}\{ \sum_{k=1}^{2} (P_{ik}Q_{mk}+P_{mk}Q_{ik})[A_{jk}(\frac{1}{z_k-c_k}) + \frac{1}{2\pi(\mu_k-\mu_\ell)} \{(\mu_\ell-\mu_k)A_{jk}J'(z_k,c_k)$$

$$+ (\mu_\ell-\bar{\mu}_\ell)\bar{A}_{j\ell}J'(z_k,\bar{c}_\ell) + (\mu_\ell-\bar{\mu}_k)\bar{A}_{jk}J'(z_k,\bar{c}_k)\}]\} \quad (j,i,m=1,2;\ \ell=3-k) \tag{6.39}$$

The constants in eq. (6.39) are defined in Chapter 5. The isotropic shear modulus is now taken to be G for clarity.

In order to evaluate the interior strain identity, (6.26), it is necessary to differentiate eq. (6.39) with respect to the cartesian coordinates of the source point ($\underset{\sim}{x}$) as follows

$$\frac{\partial}{\partial x_{h|p}} \int_A \sum_{jim} dA = 2G \text{ Re } \{ \sum_{k=1}^{2} (P_{ik}Q_{mk} + P_{mk}Q_{ik}) [A_{jk} \quad \frac{\partial}{\partial x_{h|p}} \{ \int_A \frac{\partial}{\partial z_k} \log(z_k - c_k)\, dA \} + \cdots]\} \tag{6.40}$$

The inverse of (z_k-c_k) in eq. (6.39) is written as the integral term in eq. (6.40). The nonsingular terms in eq. (6.40) are replaced by dots, as they do not contribute to the free term associated with the convected derivative, see eq. (6.27).

In order to compute the convected derivative of the Green's function domain integral in eq. (6.40), the kernel has been given in terms of the derivative of the log function which is identical to the form in eq. (6.39). Thus, we seek to evaluate the term (no sum on the index k)

$$\frac{\partial}{\partial x_{h|p}} \{ \int_A \frac{\partial}{\partial z_k} \log\,(z_k-c_k) dA \} = Q_{hk} \frac{\partial}{\partial c_k} \int_A \frac{\partial}{\partial z_k} \log(z_k-c_k) dz_k \tag{6.41}$$

In order to evaluate eq. (6.41) we begin with the Green's theorem for a real function which may be written in the form

$$\int_A \left(\frac{\partial F}{\partial x} - \frac{\partial F}{\partial y}\right) dA = \oint F(dx+dy) \tag{6.42}$$

If F(x,y) is taken as a function of $z_k, \bar{z}_k$ then

$$\begin{matrix} \partial F/\partial z_k \\ \partial F/\partial \bar{z}_k \end{matrix} = \begin{bmatrix} 1 & 1 \\ \mu_k & \bar{\mu}_k \end{bmatrix}^{-1} \begin{Bmatrix} \partial F/\partial x \\ \partial F/\partial y \end{Bmatrix} \tag{6.43}$$

with the result that

$$\frac{\partial F}{\partial z_k} = \frac{1}{\bar{\mu}_k - \mu_k} \left(\bar{\mu}_k \frac{\partial F}{\partial x} - \frac{\partial F}{\partial y}\right) \tag{6.44}$$

Taking the domain integral

$$\int_A \frac{\partial F}{\partial z_k} dxdy = \frac{1}{\bar{\mu}_k - \mu_k} \int_A (\bar{\mu}_k \frac{\partial F}{\partial x} - \frac{\partial F}{\partial y}) dA \tag{6.45}$$

we are able to use Green's theorem, (6.42), to find

$$\int_A \frac{\partial F}{\partial z_k} dA = \frac{1}{\bar{\mu}_k - \mu_k} \oint_S F d\bar{z}_k \tag{6.46}$$

Thus, eq. (6.41) may now be written as

$$Q_{hk} \frac{\partial}{\partial c_k} \int_A \frac{\partial}{\partial z_k} \log(z_k - c_k) dz_k = \frac{Q_{hk}}{\bar{\mu}_k - \mu_k} \frac{\partial}{\partial c_k} \oint_S \log(z_k - c_k) d\bar{z}_k \tag{6.47}$$

In eqs. (6.41) and (6.47), the area and path integrals are taken to exclude the small region of the singularity at $z_k = c_k$. Thus the path integral in eq. (6.47) includes the external path, a small circle surrounding c_k and an associated branch cut, as shown in Fig. 6.1.

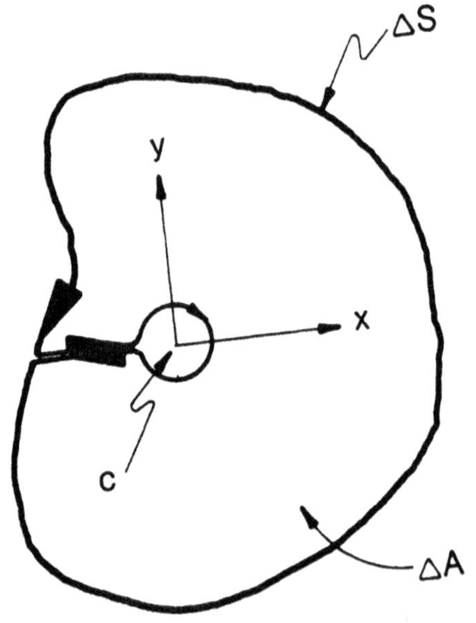

Figure 6.1. Integration path definitions

The derivative in the last term in eq. (6.47) may be formally written as (dropping the subscript for clarity)

$$\frac{\partial}{\partial c}\oint_S \log(z-c)d\bar{z} = \lim_{\Delta c\to o}\ \{\frac{1}{\Delta c}\oint_S \log(z-c-\Delta c)d\bar{z} - \oint_S \log(z-c)d\bar{z}\} \qquad (6.48)$$

This form is obtained by taking two origins for the integrals, one centered at $c(\underset{\sim}{x})$ and the other $c(\underset{\sim}{x}) + \Delta c$.

The two log terms cancel each other for the small circular paths around the singular points $c(\underset{\sim}{x})$ and $c(\underset{\sim}{x}) + \Delta c$. The integrals around the external path result in non-singular contributions given by the integral on the external path

$$-\oint_S \frac{1}{z-c}\, d\bar{z} \qquad (6.49)$$

The branch cut paths give a finite contribution to eq. (6.48). By separately taking $\Delta c = \Delta x$ and $\Delta c = \mu\Delta y$, the result is obtained

$$\frac{\partial}{\partial c}\oint_S \log(z-c)d\bar{z} = -\oint_S \frac{1}{z-c}\, d\bar{z} + 2\pi i(\bar{\mu}/\mu) \qquad (6.50)$$

The resulting terms from eqs. (6.27) and (6.40) are then given by the following constant plus integral terms

$$\frac{\partial \delta u_i}{\partial x_{h|p}} = 2G\ \mathrm{Re}\ \{\sum_{k=1}^{2} \frac{A_{ik}B_{mn}(k)}{\bar{\mu}_k - \mu_k}\ (2\pi\sqrt{-1})S_h(k)\}\delta\varepsilon^P_{mn} + \text{integral terms} \qquad (6.51)$$

where G is the isotropic shear modulus and

$$B_{mn}(k) = P_{mk}R_n(k) + P_{nk}R_m(k)\ \text{(no sum on k)} \qquad (6.52)$$

and where

$$\{R_j(k)\} = \{\begin{matrix}1\\ \mu_k\end{matrix}\}$$

$$(6.53)$$

$$\{S_j(k)\} = \{\begin{matrix}1\\ -\mu_k\end{matrix}\}$$

Taking the linear combination of displacement gradients to obtain the strain at $p(\underset{\sim}{x})$ in eq. (6.51) and defining the elastic matrix E_{ijmn} from the free term in eq. (6.51), we can write the interior strain Somigliana identity as

$$\delta\varepsilon_{ij}(p) = \int_S U^{\varepsilon}_{kij}(p,Q)\delta t_k(Q)dS + \int_S T^{\varepsilon}_{kij}(p,Q)\delta u_k(Q)dS$$
$$+ \int_A (\Sigma_{imn,j} + \Sigma_{jmn,i})\delta\varepsilon^p_{mn}(q)dA + E_{ijmn}\delta\varepsilon^p_{mn}(p) \tag{6.54}$$

with

$$\frac{1}{2G} E_{ijmn} = \mathrm{Re}\left\{ \sum_{k=1}^{2} \frac{A_{ik}B_{mn}(k)}{\bar{\mu}_k - \mu_k}(2\pi\sqrt{-1})S_j(k)\right\} \tag{6.55}$$

The bracketed result in eq. (6.55) is written in terms of the anisotropic formulation used for the fracture mechanics fundamental solution. However, the elastoplastic implementation in this section has presumed that the material is nearly isotropic. Thus, eq. (6.55) is equivalent to eq. (6.38). Should one desire to model anisotropy in the elastic material, it would be necessary to re-derive Betti's theorem in eq. (6.19) for the particular two-dimensional, plane strain case. For plane stress one would use the plane stress kernel of the fundamental solution from eq. (3.40) and derive the appropriate constant tensor for eq. (6.55).

The elastoplastic fracture mechanics formulation offers some important insight into the crack tip field behavior. Eq. (6.54) may be written in symbolic form as

$$\delta\varepsilon(\rho) = \frac{1}{\sqrt{\rho}} \langle\Delta\hat{T}^{\varepsilon}\rangle \{\delta u\} + \frac{1}{\sqrt{\rho}} \langle\Delta\hat{U}^{\varepsilon}\rangle \{\delta t\} + \langle\Delta G\rangle \{\delta\varepsilon^P\} + E\delta\varepsilon^P(\rho) \tag{6.56}$$

where ρ is the distance from the crack tip, assumed now to be small relative to the crack dimension. That the two surface integrals are inversely proportional to the square root of the distance from the crack tip derived directly from eq. (5.34) and (5.35). In order to study the behavior of the domain integral term, $\langle\Delta G\rangle$, we must study the full form for the kernel in eq. (6.54)

Assume that the plastic strains may be taken to be constant values within a set of N cells. Then

$$\int_A \Sigma_{imn,j}(p,q)\delta\epsilon^p_{mn}(q)dA = \sum_{n_c=1}^{N} \delta\epsilon^P_{mn}(n_c) \int_{\Delta A_c} \Sigma_{imn,j}dA \tag{6.57}$$

The domain integral on the cell is given by

$$\int_{\Delta A_c} \Sigma_{imn,j}(p,q)\, dA = 2\mathrm{Re}\Big\{\sum_{k=1}^{2} \Big[-A_{ik}B_{mn}(k)R_j(k)I_1(k) \tag{6.58}$$

$$+ \frac{B_{mn}(k)}{2\pi(\mu_k-\mu_\ell)} \{(\mu_\ell-\mu_k)A_{ik}R_j(k)I_2(k)$$

$$+ (\mu_\ell-\bar{\mu}_\ell)\bar{A}_{i\ell}S_j(\ell)I_3(k)$$

$$+ (\mu_\ell-\bar{\mu}_k)\bar{A}_{ik}S_j(k)I_4(k)\}\Big]\Big\} \; ; \; (\ell=3-k)$$

where

$$I_1 = \oint_{\Delta S_c} \frac{1}{z_k-c_k}\, d\bar{z}_k \tag{6.59a}$$

$$I_2 = \oint_{\Delta S_c} \frac{\partial J(z_k,c_k)}{\partial c_k}\, d\bar{z}_k \tag{6.59b}$$

$$I_3 = \oint_{\Delta S_c} \frac{\partial J(z_k,\bar{c}_\ell)}{\partial \bar{c}_\ell}\, d\bar{z}_k \tag{6.59c}$$

$$I_4 = \oint_{\Delta S_c} \frac{\partial J(z_k,\bar{c}_k)}{\partial \bar{c}_k}\, d\bar{z}_k \tag{6.59d}$$

If each cell is taken to be straight sided (e.g., triangular for each side of the cell) eq. (5.41) gives

$$B_k dS = dz_k$$

and

$$\bar{B}_k dS = d\bar{z}_k \tag{6.60}$$

Taking triangular cells with three sides for the integrals in (6.59) and dropping the k-subscripts we obtain

$$I_1 = \sum_{i=1}^{3} (\bar{B}/B)_i \; [[\log(z-c)]]_i \tag{6.61}$$

$$I_2 = \sum_{i=1}^{3} \pi(\bar{B}/B)_i \; \{[[\frac{\sqrt{(z^2-a^2)}}{\sqrt{(c^2-a^2)}}]]_i$$

$$(-1 + \frac{c}{\sqrt{(c^2-a^2)}}) \; [[\ln(z+\sqrt{(z^2-a^2)})]]_i - [[J(z,c)]]_i \tag{6.62}$$

The integrals for I_3, I_4 followed by substituting $\bar{c}_k, \bar{c}_\ell$ for c_k in eq. (6.62). The double brackets denote in eq. (6.62) taking the difference for each end of one side of the cell. From eq. (5.41) and the associated Fig. 5.2, we see that each cell side corresponds to a new value of $(\bar{B}/B)$ in eqs. (6.61) and (6.62).

Two important cases arise in the evaluation of eq. (6.56). In the first case, the integration element, ΔS_i, is taken to the limit of zero size at the crack tip with the load point c at a positive distance from the crack tip. In this case, taking

$$|z/a| = \varepsilon << 1 \tag{6.63}$$

for integration points in eq. (6.62), it is possible to show that

$$I_2 = \sum_{i=1}^{3} \pi(\bar{B}/B)_i \; O(\varepsilon)/\sqrt{(c^2-a^2)} \tag{6.64}$$

where $O(\varepsilon)$ indicates the order of magnitude of the result. Thus, the plastic strain domain integral in eqs. (6.54) and (6.56) should be convergent for plastic strain singularities up to and including $1/\varepsilon$, which is the case for perfect plasticity.

If the solution point in eq. (6.56) is taken to be within the first ring of cells around the crack tip, and the limit is taken as $\varepsilon \to o$ from (6.63), then it can be shown that the limiting terms in eq. (6.62) are of order one. For elastoplastic fracture mechanics problems, the plastic strain is singular at the crack tip. Taking the order of the singularity to be $\lambda(> 1/2)$ then eq. (6.56) has the limiting form

$$\delta\varepsilon^P = \langle\Delta G\rangle\{\delta\varepsilon^P\} + E\delta\varepsilon^P \tag{6.65}$$

Equation (6.65) is an identity for cells taken very close to the crack tip and does not provide an independent means for determining the value of λ.

If the plastic strains are non-singular, it is possible to evaluate a strain intensity factor from eq. (6.56). For non-singular plastic strains, eq. (6.56) can be evaluated for the total accumulated solution and converted into a strain intensity factor evaluation as follows

$$K^\varepsilon \propto \lim_{\rho\to o} \surd\rho\ \varepsilon(\rho) = \langle\Delta\hat{T}^\varepsilon\rangle\ \{u\}^{\cdot} + \langle\Delta\hat{U}^\varepsilon\rangle\ \{t\} + \langle\Delta\hat{G}\rangle\ \{\varepsilon^P\} \tag{6.66}$$

The first two terms on the right-hand side of eq. (6.66) have been derived in Section 5.2. The additional domain term for eq. (6.66) is taken from

$$\frac{1}{2G} \lim_{\rho\to o}(\surd\rho\textstyle\sum_{imn,j}) = \operatorname{Re}\left\{\displaystyle\sum_{k=1}^{2} \frac{\pi B_{mn}(k)}{2(\mu_k-\mu_\ell)} \frac{C_{ij}(k)}{\surd 2a} I_5(k)\right\} \tag{6.67}$$

where

$$I_5(k) = \underset{\Delta S_c}{o} \left(\frac{\bar{B}_k}{B_k}\right) \left(\frac{z_k - a - \surd(z_k^2-a^2)}{z_k-a}\right) dz_k \tag{6.68}$$

and where

$$C_{ij}(k) = (\mu_\ell-\mu_k)A_{ik}R_j(k) + (\mu_\ell-\bar{\mu}_\ell)\bar{A}_{i\ell}S_j(\ell) + (\mu_\ell-\bar{\mu}_k)\bar{A}_{ik}S_i(k) \tag{6.69}$$

Again using the triangular cell for piecewise constant plastic strains, the integral in eq. (6.68) becomes

$$I_5(k) = \sum_{i=1}^{3} (\bar{B}_k/B_k)_i\ \{\Delta z_k|_i - [[\surd(z_k^2-a^2)]]_i - a\ [[\ln(\surd(z_k^2-a^2) + z_k + a)]]\} \tag{6.70}$$

The final form of the strain intensity factor representation for nonsingular plastic strains has been shown to exist. The analytical representation is given by

$$(K_I, K_{II}) = -\int_S R_i^{I,II}(Q)u_i(Q)dS + \int_S L_i^{I,II}(Q)t_i(Q)dS + \int_A M_{ij}^{I,II}(q)\varepsilon_{ij}^P(q)dA \qquad (6.71)$$

6.6 Numerical Implementation of the Elastoplastic BIE Formulation

The elastoplastic BIE can be reduced to a numerical quadrature in the same manner as outlined for the elastic analyses in Section 3.8. The elastoplastic strains will be taken to be piecewise constant within finite two- or three- dimensional cells. Equation (6.25) can be reduced to the form of eq. (3.79)

$$[C]\ \{\delta u\} + [\Delta T]\ \{\delta u\} = [\Delta U]\ \{\delta t\} + [\Delta\Sigma]\ \{\delta\varepsilon^P\} \qquad (6.72)$$

while the interior strains identity becomes

$$\{\delta\varepsilon\} = [\Delta T^\varepsilon]\ \{\delta u\} + [\Delta U^\varepsilon]\{\delta t\} + [\Delta G]\ \{\delta\varepsilon^P\} + E\{\delta\varepsilon^P\} \qquad (6.73)$$

Hereafter, the free term at the end of eq. (6.73) will be included in the $[\Delta G]$ matrix, on the diagonal.

The solution algorithm is outlined in Fig. 6.2 where the unknown boundary data in eq. (6.72) is denoted by $\{\delta x\}$, and the known data is applied to obtain

$$[A]\ \{\delta x\} = \{\delta y\} + [\Delta\Sigma]\ \{\delta\varepsilon^P\} \qquad (6.74)$$

as in eq. (3.79) for the elastic problem.

In the first increment of loading, elastic response is assumed such that $\delta\varepsilon^P \equiv 0$ in eq. (6.72). Equation (6.73) is then used to solve for the resulting (elastic) total strain increment line in each domain cell. The total strain increment from eq. (6.74) is integrated to establish the elastoplastic stresses for each cell. Using the plain strain flow rule, (eq. 6.17), we integrate

$$\{\delta\sigma_{ij}\} = \int_{\delta\varepsilon} [C_{ijmn}^{EP}]\ \{d\varepsilon_{mn}\} \qquad (6.75)$$

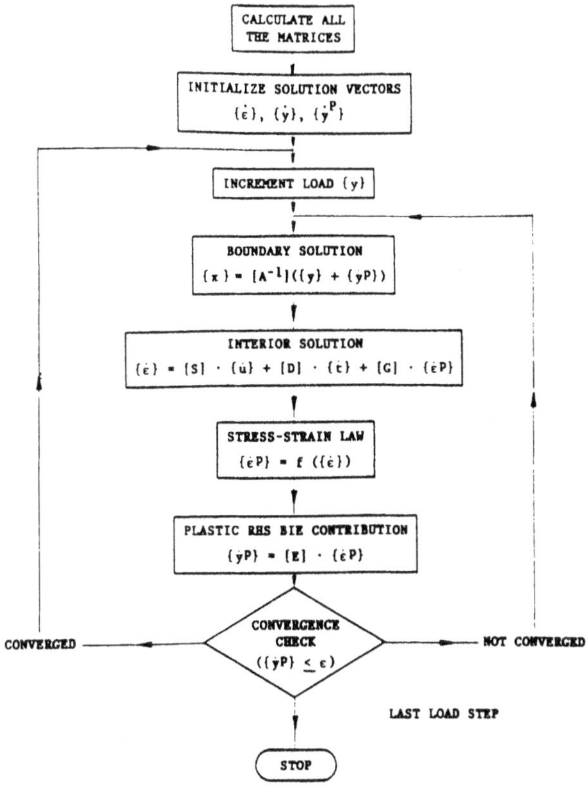

Figure 6.2. Elastoplastic solution algorithm (super-dots denote incremental changes)

where the plane-strain elastoplastic stiffness $[C^{EP}]$ is given by eq. (6.12)

$$\frac{1}{2\mu} C^{EP}_{ijmn} = [\delta_{im}\delta_{jn} + \frac{\nu}{1-2\nu}\delta_{ij}\delta_{mn} - \frac{S_{ij}S_{mn}}{2J_2(1+H/3\mu)}] \tag{6.76}$$

Eq. (6.75) is integrated by breaking the strain increment $\delta\varepsilon$ into a number of subincrements, over each of which the subincrement in total strain is taken to be constant. The number of subincrements is selected to achieve an acceptable accumulation of error in tracking the stress-strain curve of the material.

During evaluation of the stress increment from eq. (6.75), the plastic strain increment is computed from eq. (6.7). The resulting increment in plastic strain is substituted into eq. (6.74) to resolve for the increment in the boundary solution. Internal strain increments are then corrected from eq. (6.73).

As shown in Fig. 6.2, the process iterates within the increment until convergence is satisfied. Following convergence, the boundary data is incremented, and the process is repeated.

6.7 Numerical Results in Two-Dimensional Elastoplasticity

The first application problem for the elastoplastic Green's function algorithm is a center-cracked plate loaded in tension. The total width of the plate is 8 units, with a crack size of 2 units. One-quarter of the geometry was modeled using the finite element code ADINA (see Bathe (1982)), as shown in Figs. 6.3 and 6.4. Extremely fine resolution of the crack tip elements was taken in order to minimize the error in the finite element solution. The elastic stress intensity factor for this finite element model, using the crack opening displacement at the quarter-point node, was in error relative to handbook values by about 2 percent.

The BEM mesh corresponding to the local finite element scale is shown in Fig. 6.5. The elastic BEM stress intensity factor results were indistinguishable from the handbook results. The FEM/BEM meshes were selected so as to provide about three decades of plotting data in terms of crack tip distance. The maximum size of the plastic zone was limited solely for convenience in the current study.

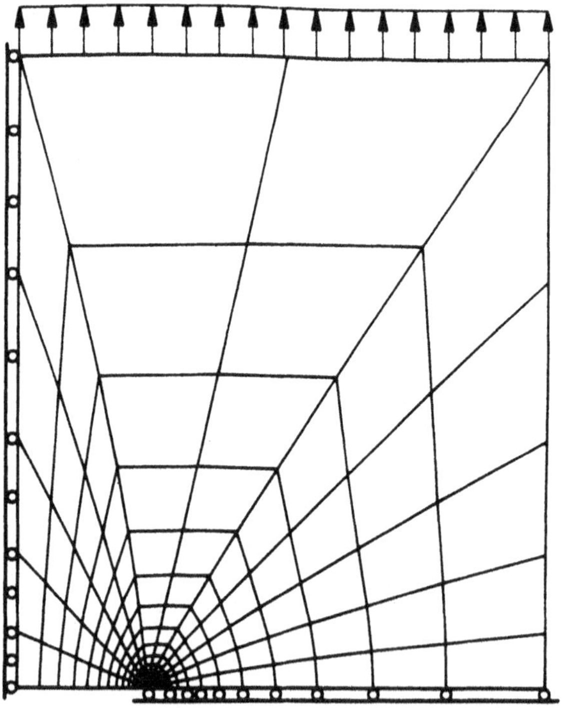

Figure 6.3. Finite-element mesh for center-cracked plate

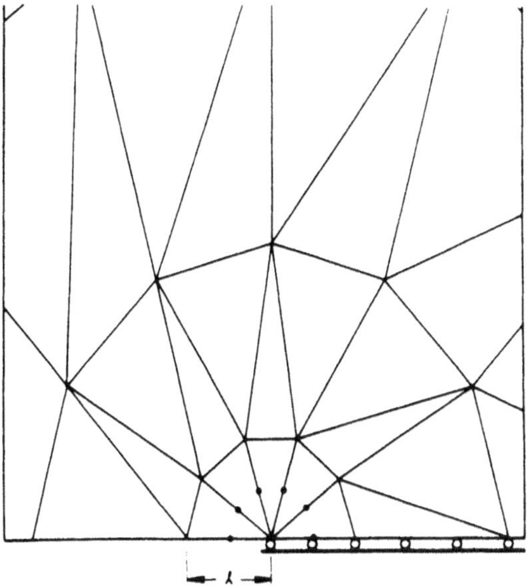

Figure 6.4. Local finite-element mesh (ℓ/a=0.001)

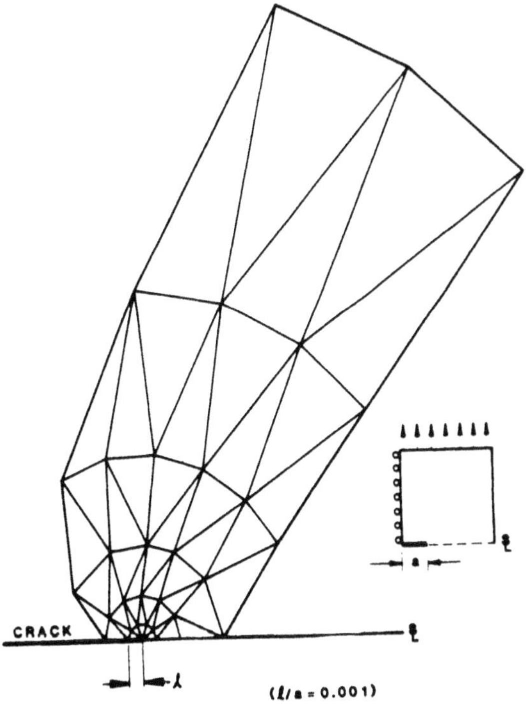

Figure 6.5. Boundary-element mesh for center-cracked plate

Plane strain conditions were used for both models. The elastoplastic material constants used were $E=2.037 \cdot 10^5 MN/m^2$, $\sigma_y=3.452 \cdot 10^2 MN/m^2$, and $\nu=0.27$.

The ADINA crack-tip model used quadratic, isoparametric finite elements with nine interior strain integration points. The BEM model used constant strain triangles throughout. As noted above, both meshes were identical in the crack tip region. The ADINA model used one layer of collapsed quadratic elements adjacent to the crack tip. This approach induces a $(1/\rho)$ type of singularity in the displacement gradient within this first layer of elements. No singularity modeling is used in the BEM plastic strain distribution.

Loading history was identical for both models and spans the range of load factors of 0.0310 to 0.2075. A value of 1.0 corresponds to yielding of the whole plate. A total of 68 load steps was used for both models. The load steps satisfy the conditions of Larsson and Carlsson (1973). Simply stated, these conditions require that at most one element becomes plastic at each load increment, and that the load increment should be smaller than 1 percent of the load corresponding to $K_{Imax}=\sigma_y\sqrt{(\pi a)}$. The range of load factors has been chosen as the range to go from yielding the innermost element to yielding the outermost element.

ADINA failed to converge for the first step until the stiffness reformulation (BFGS) procedure was used. After the first load step (requiring 20 iterations), the ADINA algorithm with reformulation generally converged with five iterations. The BEM algorithm, using elastic "stiffnesses," converged in 10 to 50 iterations at each load step, with the higher numbers occurring at the higher load levels. The total computer time for the two models was essentially the same, although the BEM calculations are cheaper per load step.

The crack-tip plastic strain distribution results are shown in Fig. 6.6 for two of the computed load levels. The data are taken from points distributed near, but not on, a line at an angle of about 85° to the plane of the crack. This angle corresponds to the line of maximum equivalent elastic strain. The jaggedness of the curves is mostly due to the use of triangular elements, as well as to the points having different angular locations. The data is plotted in terms of the centroidal value of plastic strain. The innermost row of finite elements has three sampling points radially, which accounts for the smaller radius plotted for these results. The numerical results from ADINA show a tendency for a spurious peak in plastic strain increments in the second row of elements. This peak is no doubt induced by the lack of a singularity-transition element in the current study.

It is significant to find that the numerical results are in such good agreement. This confirms the accuracy of the BEM algorithm for piecewise constant plastic strains. The BEM results do not show the strength of the plastic singularity as strongly in the first row of elements as do the finite element results, with the imbedded $1/\rho$ singularity in displacement gradient. However, both sets of results strongly indicate that the plastic strain for localized plasticity possesses the same $1/\rho$ singularity that is associated with fully developed plasticity for the case of zero strain hardening. Clearly, the presence of the underlying elastic singularity field plays an important role in enhancing the modeling accuracy for crack-tip plasticity.

The second problem was selected to validate the stress intensity factor algorithm for prior plasticity for any residual or thermal strain field. The geometry selected is a simple tension specimen with the boundary and internal mesh shown in Fig. 6.7. The mesh arrangement was selected solely for convenience, as it is used as a portion of a later mesh. The specimen was loaded to 110 percent of the yield stress for a bilinear material response. This induced a uniform plastic strain throughout the specimen.

The next step in the validation of eq. (6.66) was to introduce a crack along the bottom of the mesh as shown. The residual boundary solution corresponding to the residual internal strains is computed for the cracked case by eq. (6.74). Next, the internal strains for the residual boundary and internal variables are computed from eq. (6.73). In the case of the test problem, eq. (6.74) produced the uniform displacements compatible with uniform residual strains; eq. (6.73) computed internal strains equal to the residual strains.

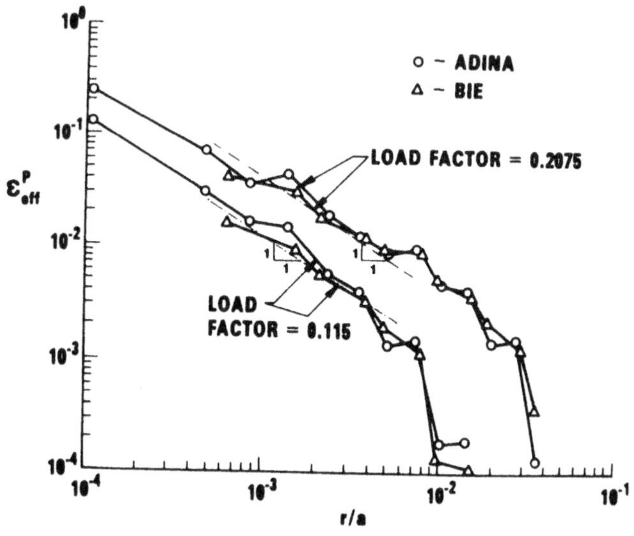

Figure 6.6. Comparison of plastic strain distributions

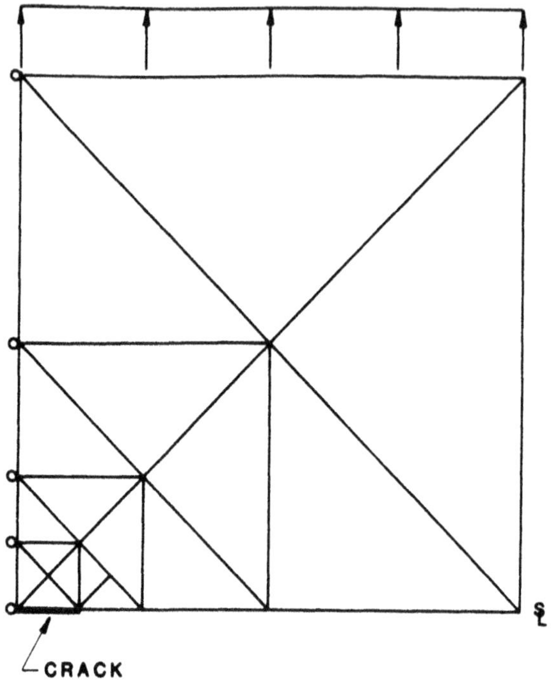

Figure 6.7. Test problem for uniform strains

The next step in the validation of eq. (6.66) was to introduce a crack along the bottom of the mesh as shown. The residual boundary solution corresponding to the residual internal strains is computed for the cracked case by eq. (6.74). Next, the internal strains for the residual boundary and internal variables are computed from eq. (6.73). In the case of the test problem, eq. (6.74) produced the uniform displacements compatible with uniform residual strains; eq. (6.73) computed internal strains equal to the residual strains.

The elastic stress intensity factor for the problem was then computed for the residual boundary terms computed from eq. (6.74). If there were further changes in the residual strains due to unloading plasticity, these would modify the elastic strain intensity factor through the appropriate term in eq. (6.66). As required for this simple case, the residual stress intensity factor was zero. The residual values in the first and third terms in eq. (6.66) cancel each other to within computer accuracy.

Figure 6.8 shows the residual stress in a large plate containing a simulated weld bead. The weld is simulated by a narrow (b/W=1/400) strip of material with plastic strains equivalent to 100 ksi elastic strains in an infinite sheet. Both plane-strain and plane-stress solutions were used. The plastic strains simulate the behavior of material which is heated to yielding in a confined region and then allowed to cool. The slight accommodation of the plate to equilibrate the residual welding stresses is seen in Fig. 6.8.

The BEM analysis was used to obtain K-solutions for various central crack sizes for cracks transverse to the weld bead. The only reason for this configuration is to be able to compare the numerical results to an exact solution for the residual stresses in Fig. 6.8.

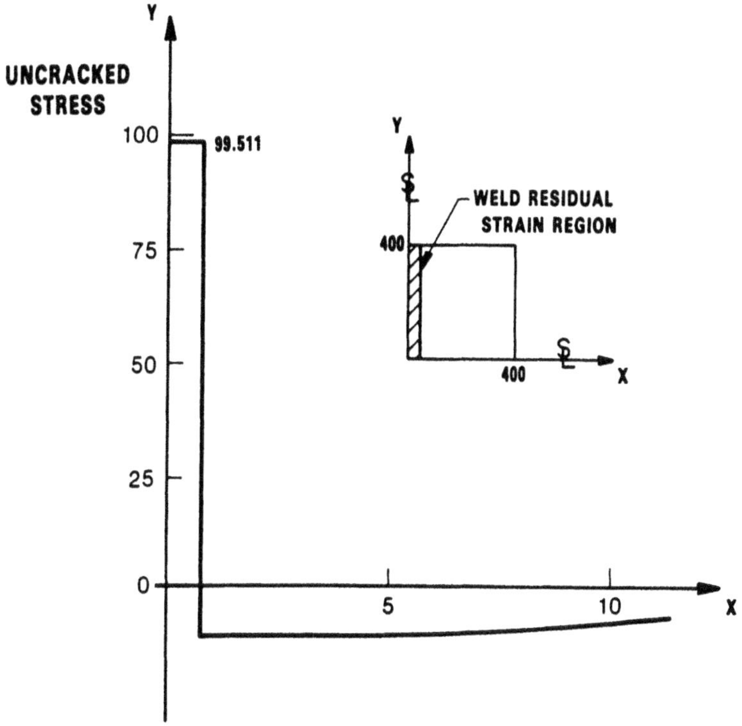

Figure 6.8. Simulated residual stresses due to welding

The approach taken is to solve eq. (6.74) to establish the boundary solution corresponding to the residual strains. Two triangles were used to integrate the plastic strains in eqs. (6.74-6.66); a single quadrilateral element would also suffice.

The results of K(a) are compared to the analytical results using an influence function approach of Tada and Paris (1983) in Fig. 6.9. The agreement is essentially exact, as expected. The volume integral in (6.66) contributes the bulk of the K-solution, as the boundary displacements associated with the weld plastic strains are very localized.

The algorithm in (6.66) is therefore seen to be a very powerful solution for residual plastic strains for geometries without known Green's functions or influence functions. Furthermore, by slight reformulations of the volume terms, other non-elastic strains, such as thermal strains or body-force strains, can be analyzed in the same manner.

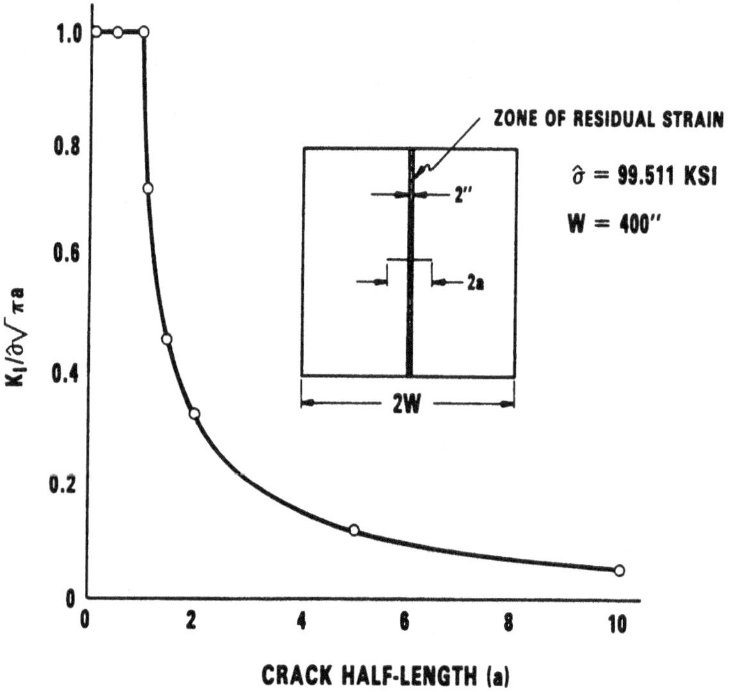

Figure 6.9. Stress intensity factor distribution for center crack in simulated welded plate

An important engineering application of this approach is the calculation of K(a) for cracks at notches (holes, fillets, etc.) which have plastic strains due to overloads or cold working. To simulate this problem, a bolt-hole specimen was modeled as shown in Fig. 6.10. The plate was loaded in plane strain to 80 percent of the net section plane-strain stress (F_{ty}=55.8 ksi). Fig. 6.11 shows the progressive development of the plastic strains, computed by the BEM program with zero crack length.

An elastic crack was then simulated for A/R=0.05, 0.50. The solution, as before, requires an equilibrium adjustment to account for the crack in the boundary solution (6.72), and a K-calculation from eq. (6.71).

Table 6.1 presents the elastic and plasticity-modified stress intensity factors for the two crack lengths. The approach used is to calculate K(a) from eq. (6.71) at no load. Table 6.1 gives the values of K(a) at no load and full load through an addition of the elastic result to the no-load solution. The results clearly demonstrate how notch plasticity causes substantial crack closure ($K(a) < 0$), thereby reducing the stress intensity factors at full load.

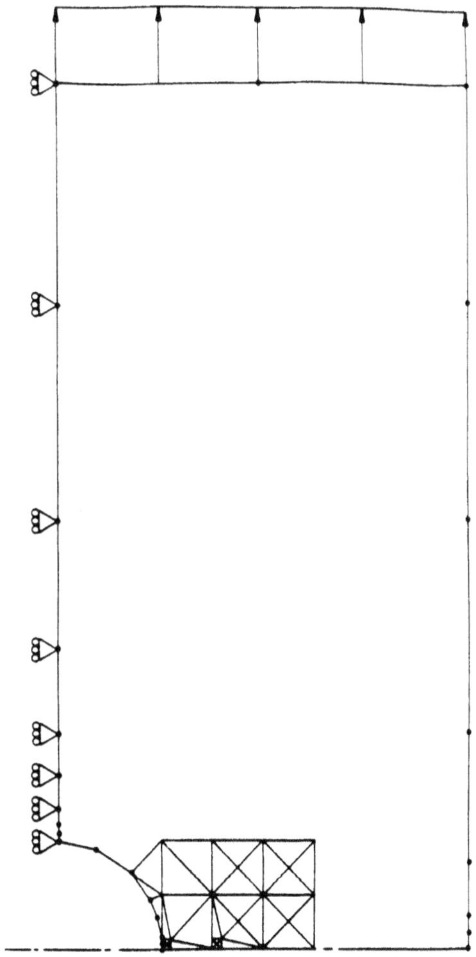

Figure 6.10. BIE model of 2:1 plate with central circular hole (R/W=0.25)

Table 6.1. Stress intensity factor results ($K_I/\sigma\sqrt{\pi a}$)

	a/R=0.05	a/R=0.5
Elastic		
-Max. load	3.411	2.070
-Zero load	0	0
Plastic		
-Max. load	0.843†	1.787†
-Zero load	-2.568‡	-0.283‡

† Maximum load values (elasticity) = elastic + zero load values.
‡ Negative values of K_I imply crack closure at positive load.

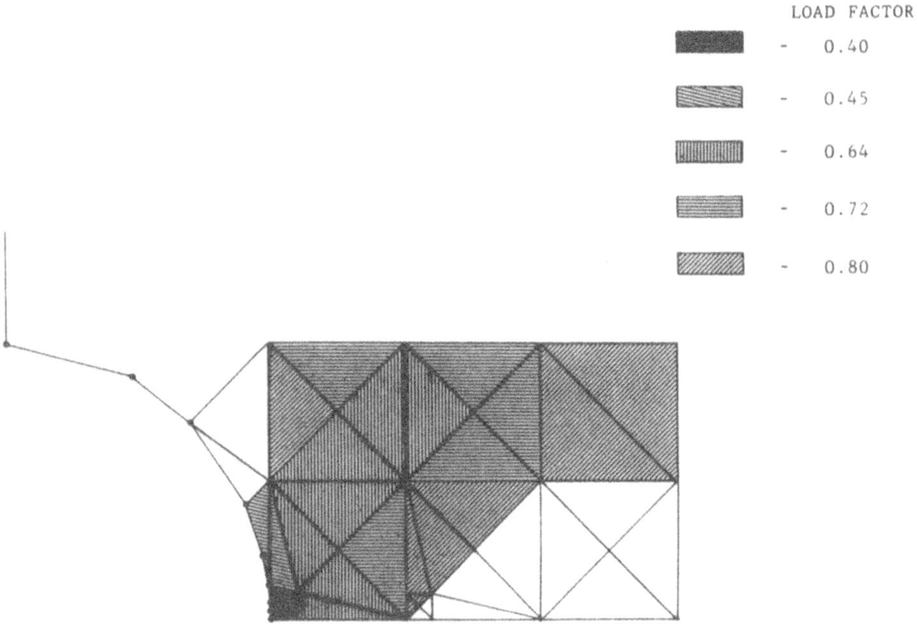

Figure 6.11. Progressive generation of plastic zone at hole (tension only)

References

(1971) J.L. Swedlow and T.A. Cruse, Formulation of boundary integral equations for three-dimensional elastoplastic flow, International Journal of Solids and Structures, 7, 1673-1683.

(1973) S.G. Larsson, and A.J. Carlsson, Influence of non-singular stress terms and specimen geometry on small-scale yielding at crack tips in elastic-plastic materials, Journal of Mechanics and Physics of Solids, 21, 263-277.

(1973) A. Mendelson, Boundary-integral methods in elasticity and plasticity, NASA TN D-7418, National Aeronautics and Space Administration, Washington, DC.

(1973) P.C. Riccardella, An implementation of the boundary-integral technique for planar problems of elasticity and elasto-plasticity, Ph.D. Thesis, Carnegie-Mellon University, Pittsburgh, PA.

(1977) S. Mukherjee, Corrected boundary-integral equations in planar thermo-elastoplasticity, International Journal of Solids and Structures, 13, 331-335.

(1978) H.D. Bui, Some remarks about the formulation of three-dimensional thermoelastoplastic problems by integral equations, International Journal of Solids and Structures, 14, 935-939.

(1981) M. Morjaria and S. Mukherjee, Numerical Analysis of Planar, Time-Dependent Inelastic Deformation of Plates with Cracks by the Boundary Element Method, International Journal of Solids and Structures, 17, 127-143.

(1982) S. Mukherjee, Boundary Element Methods in Creep and Fracture, Applied Science, London.

(1983) J.C.F. Telles, The Boundary Element Method Applied to Inelastic Problems, Springer-Verlag, Berlin.

(1986) T.A. Cruse and E.Z. Polch, Elastoplastic BIE Analysis of Cracked Plates and Related Problems, Part 1: Formulation, International Journal for Numerical Methods in Engineering, 23, 429-437.

(1986) T.A. Cruse and E.Z. Polch, Elastoplastic BIE Analysis of Cracked Plates and Related Problems, Part 2: Numerical Results, International Journal for Numerical Methods in Engineering, 23, 439-452.

7.0 Displacement Discontinuity Modeling of Cracks

7.1 Introduction

The limitations inherent to the BIE formulation of fracture mechanics problems were outlined in Chapter 4. It was shown that the integral equations degenerate when two surfaces of the same region become arbitrarily close.

Two methods for dealing with the degeneration of the BIE fracture mechanics formulation have been presented. Chapter 4 presents the multiregion implementation wherein the body is divided into at least two regions, each containing one of the crack surfaces. Each subregion now has a regular surface and contains a well-posed elasticity problem. The subregions are joined at mutual surfaces through continuity relations.

While the solution technique using multiple regions provides quite satisfactory numerical results, the dividing surface between the subregions is arbitrary, except for the crack itself. Further, the use of surfaces on the inside of the continuum has some degrading effect on the accuracy of the solution. Finally, for fatigue crack growth modeling, crack extension following some growth criteria will likely result in the need for a full internal remeshing to keep the crack surface a part of the subregion boundaries.

The second method for eliminating the problem of the BIE degeneration is the Green's function method presented in Chapter 5. While the Green's function can, in theory, be applied to a wide range of crack shapes, the method is limited to two dimensions and requires considerable extra programming effort. The accuracy of the Green's function solutions is essentially superior to any numerical approach strictly, as the crack field is exactly contained in the internal Somigliana identities.

A third method exists, and this is referred to herein as the displacement discontinuity method. In this formulation the crack is directly treated as a single surface across which the displacements are discontinuous. An integral equation written in terms of the applied crack surface fractions can then be formulated and solved numerically. While still an area of research, the displacement discontinuity method deserves review in this chapter. Further research progress in this area is needed and is, in fact, in various stages of the archival publication process.

7.2 Formulation of the Three-Dimensional Traction BIE for Flat Cracks

The initial formulation steps for the flat crack problem are outlined in Section 4.2. The crack surface geometry is defined in terms of an upper (+) and lower (-) surface, as shown in Fig. 7.1. For the case of equal applied tractions on the two crack surfaces, eq. (4.2) gives the Somigliana identity for internal displacements

$$u_i(p) = - \int_{\Gamma} T_{ij}(p,Q)\ \Delta u_j(Q)\ dS \tag{7.1}$$

where

$$\Delta u_j = u_j(Q^+) - u_j(Q^-) \tag{7.2}$$

In the case of an additional surface **S**, the term in eq. (7.1) must be augmented by the full integrals on **S** given in eq. (4.1).

As outlined in Section 3.6, the internal stress at the source point $p(\underset{\sim}{x})$ is obtained by differentiation of eq. (7.1) resulting in

$$\sigma_{ij}(p) = - \int_{\Gamma} T^{\sigma}_{kij}\ (p,Q)\ \Delta u_k(Q)\ dS \tag{7.3}$$

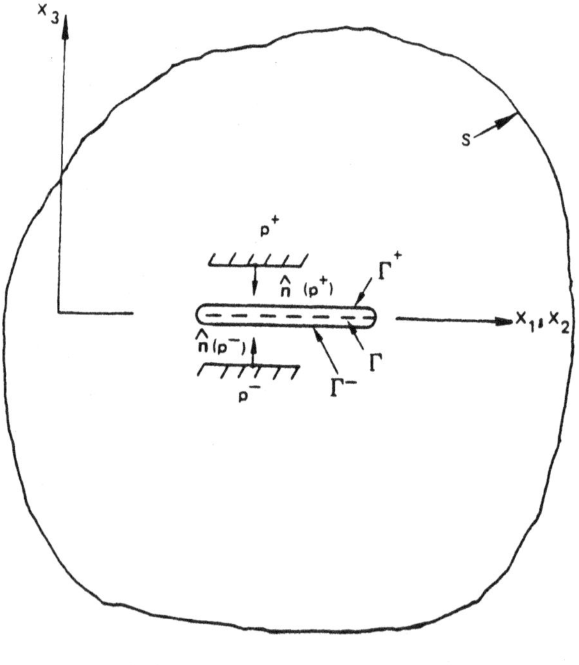

Figure 7.1. Definition of flat crack modeling terms

where

$$8\pi\, T^{\sigma}_{kij}(p,Q)/\mu = \frac{2\nu}{1-\nu}\left(\delta_{ij}\nabla^2 r,_{k3} + \delta_{k3}\nabla^2 r,_{ij}\right) + \delta_{kj}\nabla^2 r,_{3i} + \delta_{ki}\nabla^2 r,_{3j}$$

$$+ \delta_{3j}\nabla^2 r,_{ki} + \delta_{3i}\nabla^2 r,_{kj} - \frac{2}{1-\nu}\, r,_{ijk3} \tag{7.4}$$

Equation (7.4) is written in a form that is convenient for the geometry shown in Fig. 7.1, where x_3 is the coordinate normal to the crack surface. Note the change of coordinates from Fig. 4.1 made for current convenience. The distance between the source point $p(\tilde{x})$ and the field point $q(\tilde{x})$ or $Q(\tilde{x})$ is given by the usual distance term

$$r^2 = \sum_{i=1}^{3} (x_{ip}-x_{iQ})^2 \tag{7.5}$$

Equation (7.3) is generally referred to as hyper-singular in that $p(\tilde{x}) \to P(\tilde{x})$ results in an unbounded result for $r \to o$. The hyper-singularity will be removed through an integration-by-parts of eq. (7.3). The following three-dimensional identities are useful

$$\nabla^4 r(p,Q) = 0 \qquad p \neq Q \tag{7.6}$$

$$r,_{ij} = \left(\delta_{ij}-r,_i r,_j\right)/r(p,Q) \tag{7.7}$$

$$\nabla^2 r(p,Q) = 2/r(p,Q) \tag{7.8}$$

Combining eq. (7.8) and eq. (7.7), we find that

$$\nabla^2 r_{,ij} = -\,8\left(\delta_{ij}-r_{,i}r_{,j}\right)/r \tag{7.9}$$

Three traction components constitute loading on the crack surface. These tractions are given for the global coordinate system in Fig. 7.1, as the negative of the stresses

$$t_i(P) = -\,\sigma_{i3}(P) \tag{7.10}$$

Letting two internal points (p^+,p^-) be selected as shown in Fig. 7.1, the stresses from eq. (7.3) are given as

$$\sigma_{i3}(p^{\pm}) = -\frac{\mu}{8\pi}\left\{\int_{\Gamma}\left[\frac{1+\nu}{1-\nu}\left(\delta_{i3}r_{,k3} + \delta_{k3}r_{,i3}\right)\right.\right.$$

$$\left.\left. + \delta_{ki}\nabla^2 r_{,33} + \nabla^2 r_{,ki} - \frac{2}{1-\nu}r_{,ik33}\right]\Delta u_k(Q)\,dS\right\} \tag{7.11}$$

Integration-by-parts of eq. (7.11) can be performed for derivatives in the x_1, x_2 plane. Letting the Greek subscripts $\alpha, \beta = 1,2$ refer to these in-plane directions, the following identity will be used

$$\int_{\Gamma}\frac{\partial f}{\partial x_\alpha}\Delta u_k\,dS = \oint_C f\,\Delta u_k dx_\alpha - \int_{\Gamma'} f\,\frac{\partial \Delta u_k}{\partial x_\alpha}\,dS \tag{7.12}$$

Using the properties of the operators in eq. (7.6) to (7.9), we write

$$r_{,ik33} = \nabla^2 r_{,ik} - r_{,ik11} - r_{,ik22} \tag{7.13}$$

Equation (7.13) can be used in eq. (7.11) to eliminate nonintegrable $r_{,3333}$ terms and odd-order derivatives of $r(p,Q)$ with respect to x_3. It is then possible to add the $\sigma_{i3}(p^+)$ and $\sigma_{i3}(p^-)$ terms to obtain the following traction terms

$$\sigma^0_{13}(p^+) = \left[\sigma^0_{13}(p^+) + \sigma^0_{13}(p^-)\right]/2 = \frac{\mu}{4\pi(1-\nu)}\left\{\int_{\Gamma}\left[(\Delta u_{1,1} + \Delta u_{2,2})\frac{r_{,1}}{r^2}\right.\right.$$

$$\left. + (1-\nu)(\Delta u_{1,2} - \Delta u_{2,1})\frac{r_{,2}}{r^2}\right]dS$$

$$- \oint_C\left[\left(\Delta u_1\frac{r_{,1}}{r^2} - (1-\nu)\Delta u_2\frac{r_{,2}}{r^2}\right)dx_1\right. \tag{7.14}$$

$$\left.\left. + \left(\Delta u_2\frac{r_{,1}}{r^2} + (1-\nu)\Delta u_1\frac{r_{,2}}{r^2}\right)dx_2\right]\right\}$$

$$\sigma_{23}^{0}(p^{+}) = \frac{\mu}{4\pi(1-\nu)} \{\int_{\Gamma} [(\Delta u_{1,1} + \Delta u_{2,2}) \frac{r_{,2}}{r^2} + (1 - \nu)(\Delta u_{2,1} - \Delta u_{1,2}) \frac{r_{,1}}{r^2}] dS - \oint_{C} [(\Delta u_2 \frac{r_{,2}}{r^2} - (1 - \nu)\Delta u_1 \frac{r_{,1}}{r^2}) dx_2 + (\Delta u_1 \frac{r_{,2}}{r^2} + (1 - \nu)\Delta u_2 \frac{r_{,1}}{r^2}) dx_1]\} \tag{7.15}$$

$$\sigma_{33}^{0}(p^{+}) = \frac{\mu}{4\pi(1-\nu)} \{\int_{\Gamma} [\Delta u_{3,1} \frac{r_{,1}}{r^2} + \Delta u_{3,2} \frac{r_{,2}}{r^2}] dS - \oint_{C} [\Delta u_3 \frac{r_{,1}}{r^2} dx_1 + \Delta u_3 \frac{r_{,2}}{r^2} dx_2]\} \tag{7.16}$$

Equations (7.14) and (7.15) are coupled in-plane responses for the shear-loaded crack. As will be shown in Section 7.4, these responses correspond to the coupled Mode II and Mode III responses of three-dimensional cracks. The third traction BIE, (7.16), is the uncoupled Mode I response. All three equations are mathematically equivalent for numerical analysis. Equations (7.14) to (7.16) apply to the case of equal tractions on the upper and lower crack surfaces as $p(\underset{\sim}{x}) \to P(\underset{\sim}{x})$.

Limiting forms of eqs. (7.14) to (7.16) without the hyper-singularity at $r(P,Q)=0$ are obtained if it is assumed that the displacement derivatives are continuous at $P(\underset{\sim}{x})$. In that case the integrals may be evaluated for a small circular disk Γ_ε centered at $P(\underset{\sim}{x})$ as follows

$$\int_{\Gamma_\varepsilon} \Delta u_{i,\alpha}(Q) \frac{r_{,\beta}}{r^2} dS = \Delta u_{i,\alpha}(P) \int_{\Gamma_\varepsilon} (\frac{r_{,\beta}}{r^2}) dS + \int_{\Gamma_\varepsilon} [\Delta u_{i,\alpha}(Q) - \Delta u_{i,a}(P)] (\frac{r_{,\beta}}{r^2}) dS \tag{7.17}$$

The first term on the right-hand side of eq. (7.17) is easily seen to be zero due to the odd order trigonometric term $r_{,\beta}$. The second term has zero limiting value for continuity, as the square bracket is of order ε and the area is of order ε^2. Thus, the integral over Γ_ε is zero, and the principal value of eqs. (7.14) to (7.16) exists as $p(\underset{\sim}{x}) \to P(\underset{\sim}{x})$. Note that the strength of the principal value interpretation relies on the odd-trigonometric term in eq. (7.17); this is much weaker than for the usual BIE discussed in Section 3.7. The weakness has much to do with the problem of obtaining direct numerical solutions to eqs. (7.14) to (7.16).

Then, taking the principal values of the integrals in eqs. (7.14) to (7.16), we are able to replace $p^+(\underset{\sim}{x})$ by $P^+(\underset{\sim}{x})$ to obtain the traction BIE for flat crack analysis.

7.3 Formulation of the Two-Dimensional Traction BIE

Let the surface of discontinuity be as shown in Fig. 4.1, with x_2 normal to the crack plane. Also consider the problem to be limited to plane stress or plane strain such that the Mode III response is zero. The Mode I and Mode II responses are then uncoupled and have the same mathematical form. Only the Mode I formulation is considered herein.

The Galerkin vector for the two-dimensional fundamental solution is given from eq. (3.10) as

$$G_i^* = r^2(1-\log r)e_i/8\pi\mu \tag{7.18}$$

and the displacements as

$$u_i^* = \nabla^2 G_i^* - G_{j,ij}^*/2(1-\nu) \tag{7.19}$$

where

$$\nabla^2 = \partial^2/\partial r^2 + (1/r)\partial/\partial r \tag{7.20}$$

The interior displacement Somigliana identity for the case $\Delta u_1 \equiv 0$ is given by

$$u_m(p) = \mu \int_\Gamma \left[\frac{\nu}{1-\nu}\nabla^2 G_{,m} + 2\delta_{m2}\nabla^2 G_{,2} - \frac{1}{1-\nu} G_{,m22}\right] \Delta u_2 \, dS \tag{7.21}$$

where we can replace $G_{,m22}$ for integration as

$$G_{,m22} = \nabla^2 G_{,m} - G_{,m11} \tag{7.22}$$

The appropriate interior stress term for the Mode I traction BIE is given from Hooke's law as

$$\sigma_{22}/\mu = \frac{2\nu}{1-2\nu} u_{m,m} + 2u_{2,2} \tag{7.23}$$

such that the Somigliana identity for stress is given by

$$\sigma_{22}/2\mu^2 = -\int_\Gamma [-\frac{1}{1-\nu} \nabla^2 G_{,11} + \frac{1}{1-\nu} G_{,2211}] \Delta u_2 \, dS \tag{7.24}$$

Taking the x_1-derivative for integration by parts we obtain

$$\sigma_{22}/2\mu^2 = \frac{1}{1-\nu} \{[\nabla^2 G - G_{,22}]_{,1} \Delta u_2 |_{x_{1}=+a}^{x_1=-a} - \int_\Gamma [\nabla^2 G_{,1} - G_{,221}] \Delta u_{2,1} \, dS\} \tag{7.25}$$

which is equivalent to eq. (7.16).

The terms in eq. (7.18) may be substituted into eq. (7.25) to obtain the needed form of the Somigliana identity

$$2(1-\nu)\pi \, \sigma_{22}/\mu = \int_\Gamma \frac{r_{,1}(1+r_{,2}r_{,2}+r_{,2})}{r} \Delta u_{2,1} \, dS \tag{7.26}$$

As in the three-dimensional case, eq. (7.26) is taken for p^+ and p^-, and the limiting form of the sum of the two equations is obtained for $p(\tilde{x}) \to P(\tilde{x})$. The traction loading on Γ is again taken to be equal on both crack surfaces such that the traction BIE is obtained

$$2(1-\nu)\pi \, \sigma_{22}(x_1)/\mu = -\int_{-a}^{a} \frac{\Delta u_{2,1}}{x_1-\xi_1} d\xi_1 \tag{7.27}$$

As in the three-dimensional case, the traction BIE is to be interpreted in the principal value sense.

Using the full two-dimensional equations, it is equally easy to derive the uncoupled shear loading traction BIE as follows

$$2(1-\nu)\pi \, \sigma_{12}(x_1)/\mu = -\int_{-a}^{a} \frac{\Delta u_{1,1}}{x_1-\xi_1} d\xi_1 \tag{7.28}$$

The two-dimensional traction BIE, (7.27), has an analytical solution, as shown by Kanwal (1971) to be

$$\frac{\mu}{1-\nu} \frac{\partial u_2}{\partial x_1} = -\frac{1}{\pi} \left(\frac{a+x_1}{a-x_1}\right)^{1/2} \int_{-a}^{a} \left(\frac{a-\xi_1}{a+\xi_1}\right)^{1/2} \frac{\sigma_{22}(\xi_1) d\xi_1}{\xi_1 - x_1} - iC/\pi\sqrt{(a^2-x_1^2)} \tag{7.29}$$

The constant of integration C in eq. (7.29) is used to eliminate singular terms from integrating the loading in eq. (7.29).

As an example, take the problem of a uniform crack surface traction $\sigma_{22}(\xi_1)=-p$. Then, integrating eq. (7.29) we obtain

$$\frac{\pi\mu}{1-\nu}\frac{\partial u_2}{\partial x_1} = p\sqrt{(a^2-x_1^2)}\int_{-a}^{a}\frac{d\xi_1}{\sqrt{(a^2-\xi_1^2)}(\xi_1-x_1)} - \frac{p(a+x)\pi}{\sqrt{(a^2-x_1^2)}} - \frac{iC}{\sqrt{(a^2-x_1^2)}} \qquad (7.30)$$

Taking $C=p\pi ai$, the last two terms in eq. (7.30) cancel, and the principal value integral is evaluated with the final result given by

$$\mu\partial u_2/\partial x_1/(1-\nu) = -\, px_1/\sqrt{(a^2-x_1^2)} \qquad (7.31)$$

which may be integrated to obtain

$$\mu\, u_2/(1-\nu) = p\sqrt{(a^2-x_1^2)} \qquad (7.32)$$

7.4 Near Crack Tip Solution to BIE

The traction BIE formulation from eq. (7.14) to eq. (7.16) forms the basis of a study of the three-dimensional crack front stresses. The point p(x) will be taken in the plane of the crack Γ, taken for now to be $x_3=0$, and a slight distance δ ahead of the crack front. Let x_2 be normal to the crack front, directed along the crack surface, and x_1 along the crack front. The relative crack surface motions near the crack tip are represented asymptotically by

$$\Delta u_i = 2C_i(x_1)x_2^{\lambda} \qquad (7.33)$$

where $C_i(x_1)$ is a smooth function, and $\lambda > 0$.

The in-plane gradients of the relative crack surface motion are given by

$$\begin{aligned}\Delta u_{i,1} &= 2(dC_i/dx_1)x_2^{\lambda}\\ \Delta u_{i,2} &= 2\lambda C_i(x_1)x_2^{\lambda-1}\end{aligned} \qquad (7.34)$$

Let $p(x)$ be located at $(x_1, -\delta, 0)$, where δ is taken to be very small relative to the in-plane dimensions of the crack, and let $r(p,Q)$ be the Cartesian distance between $p(x)$ and the integration point. Then eq. (7.16) can be rewritten for the case of Mode I response the limit as $\delta \to 0$

$$\sigma_{33}(p) = \frac{\mu}{4\pi(1-\nu)} \int_{-\infty}^{\infty} d(\frac{x_1}{\delta}) \int_{0}^{\infty} \Delta u_{3,\alpha} \frac{r,a}{(r/\delta)^2} d(\frac{x_2}{\delta}) \tag{7.35}$$

Now let $\xi = x_1/\delta$ and $\eta = x_2/\delta$ such that $(r/\delta)^2 = (1+\eta)^2 + \xi^2$. It is easily established that the term for $\Delta u_{3,1}$ gives a zero result in eq. (7.20). Taking then the crack opening term in eq. (7.19) with $r,_1 = (1+\eta)/(r/\delta)$, eq. (7.20) becomes

$$\sigma_{33}(p) = \frac{\mu\delta^{\lambda-1}}{4\pi(1-\nu)} \int_{-\infty}^{\infty} d\xi \int_{0}^{\infty} 2\lambda C_3 \eta^{\lambda-1} \frac{(1+\eta)d\eta}{[(1+\eta)^2+\xi^2]^{3/2}} \tag{7.36}$$

In the local vicinity of δ, $C_3(\xi)$ is approximately constant such that the first integral can be evaluated

$$\sigma_{33}(p) = \frac{\mu\delta^{\lambda-1}}{4\pi(1-\nu)}(2\lambda C_3)2 \int_{0}^{\infty} \frac{\eta^{\lambda-1}d\eta}{\eta+1} \tag{7.37}$$

$$= \frac{\mu\lambda\delta^{\lambda-1}C_3}{1-\nu} \text{ Cosecant}[(1-\lambda)\pi] \tag{7.38}$$

It can be established separately, for σ_{33} to be finite and real on the crack surface, that $\lambda = 1/2$, with the following result

$$\sigma_{33}(p) = \frac{\mu}{1-\nu} (\frac{C_3}{2}) \frac{1}{\sqrt{\delta}} \tag{7.39}$$

In the case of the two-dimensional crack opening displacement solution

$$C_3 = 2(1-\nu)K_I/\mu\sqrt{(2\pi)} \tag{7.40}$$

Use of eq. (7.39) in eq. (7.40) results in the crack tip stress for the three-dimensional problem

$$\sigma_{33}(p) = K_I/\sqrt{(2\pi\delta)} \tag{7.41}$$

which, of course, is the same as the two-dimensional result.

Thus, it is seen that for any interior point p(x) the stress state ahead of the three-dimensional crack has a $1/\sqrt{\delta}$ singularity identical to the two-dimensional results. In fact, if the crack surface is only taken to be from zero to infinity (corresponding to p(x) at the surface) then the result is still that for the two-dimensional case, eq. (7.25), except for a division by a factor of 2. By inversion, finite crack surface applied stresses must lead to a square-root singular stress (strain) field ahead of the crack.

7.5 Current Numerical Method

The current numerical method uses a finite element interpolation scheme for the unknown displacement discontinuities at the crack surface and their derivatives. In the following, the term "displacements" is used for "displacement discontinuities" for brevity. By themselves, boundary element interpolations produce undesirable discontinuities of displacement derivatives at boundaries of the elements; thus, higher order derivatives are used as nodal quantities.

A special interpolation procedure is used to satisfy continuity requirements for the displacement gradients. Displacement fields on the crack surface are modeled using quadratic 8-noded isoparametric elements (with quarter-point modeling of $O(\sqrt{R})$ displacement variation) to the crack front. The same type of interpolation is used for displacement gradients, except at the crack front, where the explicit crack front singularity of $O(\frac{1}{\sqrt{R}})$ (where R is the distance from the crack front) is introduced. The relationship between these two interpolations is established by the following schemes.

The interpolation of the displacements on a crack is given by the relation

$$V_i(P) = \Delta u_i(P) = N^k(P)\Delta u_i^k = N^k(P) \cdot V_i^k \qquad (7.42)$$

where

i = 1,3
k = 1,n
P = point on the crack surface
n = number of nodes of the mesh
$N^k(P)$ = shape function for node number k (quadratic serendipity shape function)

Differentiation of $V_i(P)$ with respect to global variable x_α will provide the in-plane displacement gradient field

$$V_{i,\alpha}(P) = \frac{\partial V_i}{\partial x_\alpha}(P) = \frac{\partial N^k}{\partial x_\alpha}(P) \cdot V_i^{\,k} \qquad (\alpha=1,2) \tag{7.43}$$

This field will be, in general, discontinuous at the element boundaries as are the stresses and strains in standard finite elements.

To find a continuous displacement gradient field corresponding to the original displacement field, the same mesh is used to interpolate displacement gradients and to establish a relationship between the set of nodal displacements and the set of nodal displacement gradients. The interpolation of displacement gradients will be given by the relation

$$V_{i,\alpha}(P) = M^j(P) V_{i\ ,\alpha}^{\,j} \tag{7.44}$$

where

$j \quad = 1,m$

$M^j(P)$ = shape function for mode number j (quadratic serendipity function for noncrack front elements; explicit singularity serendipity function for elements adjacent to the crack front).

The nodal values of displacement gradients are found by minimizing the difference between the two interpolations, (7.43) and (7.44). The difference between the interpolations on the whole crack surface (indices k,j running through all nodes of the crack model) can be written in shorter form as $E_{i\alpha}$

$$E_{i\alpha}(P) = \frac{\partial N^k}{\partial x_\alpha}(P) V_i^{\,k} - M^j(P) V_{i\ ,\alpha}^{\,j} \qquad \begin{array}{l} k=1,n \\ j=1,n \\ i=1,3 \\ \alpha=1,2 \end{array} \tag{7.45}$$

Using the least-squares procedure, we obtain

$$\partial\left(\int_\Gamma E_{i\alpha}^2(P)\,dS\right) / \partial V_{i\ ,\alpha}^{\,r} = 0 \qquad \begin{array}{l} r=1,n \\ i=1,3 \\ \alpha=1,2 \end{array} \tag{7.46}$$

Substitution of eq. (7.44) results in

$$\int_\Gamma \left(\frac{\partial N^k}{\partial x_\alpha} V_i^{\,k} M^r - M^j V_{i\ ,\alpha}^{\,j} M^r \right) dS = 0 \qquad r=1,n \tag{7.47}$$

or

$$\int_\Gamma M^j M^r dS \cdot V_{i,\alpha}^{\;j} = \int_\Gamma \frac{\partial N^k}{\partial x_\alpha} M^r dS \cdot V_i^{\;k} \qquad r=1,n \tag{7.48}$$

The matrix representation of eq. (7.48) takes on the following form

$$[MM]\{V_{i,\alpha}\} = [NM]\ \{V_i\} \tag{7.49}$$

where

$[MM]$ = square matrix of dimensions 6n x 6n

with $MM_{jr} = \int_\Gamma M^j M^r dS$

$[NM]$ = rectangular matrix of dimensions 6n x 3n

with $NM_{kr\alpha} = \int_\Gamma \frac{\partial N^k}{\partial x_\alpha} M^r dS$

$\{V_{i,\alpha}\}$ = column vector of nodal displacement gradients (dimension 6n x 1)

$\{V_i\}$ = column vector of nodal displacements (dimension 3n x 1)

The above relationships hold for all the nodes except those on the crack front. Inspection of the integrals in (7.48) reveals that for the nodes lying on the crack front, the integrals are unbounded. Since $M^j(P) \sim O(\frac{1}{\sqrt{R}})$ because of the explicit displacement gradient singularity, and $\frac{\partial N^j}{\partial x_\alpha} \sim O(\frac{1}{\sqrt{R}})$ because of the implicit singularity of displacement gradients obtained from quarter-point elements, we have

$$\int_\Gamma M^j M^j dS = \int_\Gamma O(\frac{1}{\sqrt{R}}) O(\frac{1}{\sqrt{R}}) dS = \int_\Gamma O(\frac{1}{R}) dS = \infty \tag{7.50}$$

as well as

$$\int_\Gamma \frac{\partial N^j}{\partial x_\alpha} M^j dS = \int_\Gamma O(\frac{1}{\sqrt{R}}) O(\frac{1}{\sqrt{R}}) dS = \int_\Gamma O(\frac{1}{R}) dS = \infty \tag{7.51}$$

for subscript j corresponding to the nodes on the crack front. To remedy this problem, we rewrite eq. (7.47)

$$\int_{\Gamma_r} M^r\left(\frac{\partial N^k}{\partial x_\alpha} \cdot V_i^{\ k} - M^j V_{i\ ,\alpha}^{\ j}\right) dS = 0 \tag{7.52}$$

for r corresponding to the nodes on the crack front and the domain of integration Γ_r being a patch of elements containing node number r (the area on which shape function M^r is different from zero). In order for integral (7.52) to vanish, we may impose

$$\frac{\partial N^k}{\partial x_\alpha} V_i^{\ k} = M^j V_{i\ ,\alpha}^{\ j} \tag{7.53}$$

on Γ_r (that is for all subscripts k and j corresponding to the nodes of the element patch Γ_r of Fig. 7.2).

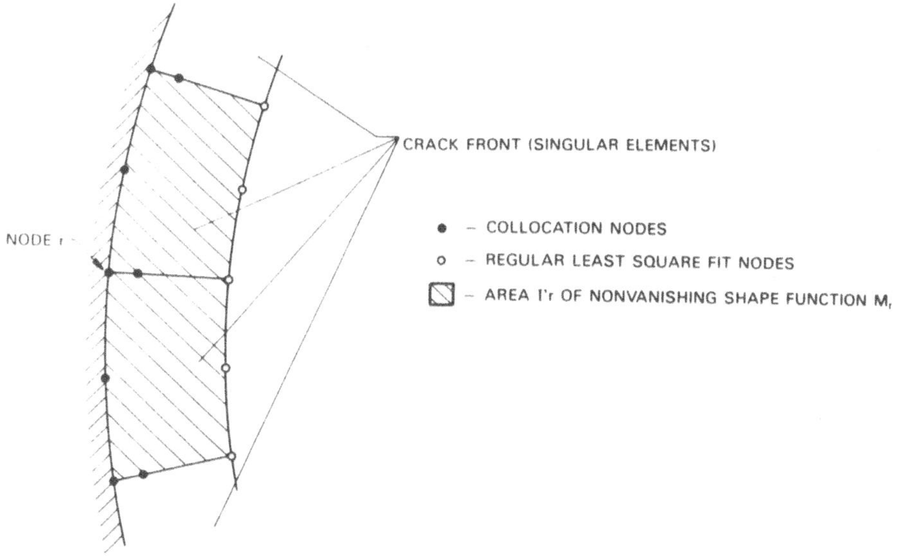

Figure 7.2. Crack front modeling

If the terms on both sides of eq. (7.53) have the same expansion in terms of in-plane coordinates, then eq. (7.53) can be satisfied exactly by matching the coefficients of these expansions. The match of the coefficients can be accomplished by collocating eq. (7.53) at the same number of points as the number of terms in the expansion. For example, in a one-dimensional case, we have Henshell and Shaw (1975) (R is the crack tip distance):

$$\frac{\partial N^k}{\partial x_\alpha} V_i^k = A_1/\sqrt{R} + B_1, \qquad (7.54)$$

and if shape functions M^j are selected such that

$$M^j V_{i,\alpha}^j = A_2/\sqrt{R} + B_2, \qquad (7.55)$$

it is possible to match the coefficients A_1 to A_2 and B_1 to B_2 by collocating at two points on an element. In the case considered (two dimensions) the exact match is impossible because of the presence of certain different terms on the left- and right-hand sides of eq. (7.53). However, the match of the principal terms $O(R^{-1/2})$ and $O(1)$ (where R is the local crack front distance) between left and right sides of eq. (7.53), obtained by collocating the equation at five points of each crack front element (three nodes at the crack front and two quarter points, Fig. 7.2), gives a very good match of displacement gradients.

In the computer implementation, the rows of matrix eq. (7.50) corresponding to the nodes on the crack front as well as the quarter points are replaced by the matrix form of eq. (7.53)

$$[N]\,\{V_i\} = [M]\{V_{i,\alpha}\} \qquad \begin{matrix} i=1,3 \\ \alpha=1,2 \end{matrix} \qquad (7.56)$$

evaluated at the collocation points. The positioning of collocation points corresponding to the crack front nodes takes into account the loss of singularity for the curved serendipity quarter-point element.

The matrix MM of eq. (7.49) has the same structure as the finite element consistent mass matrix; clearly, MM is positive definite. Nodal values of the displacement gradients are then calculated by

$$\{V_{i,\alpha}\} = [MM^{-1}][NM]\{V_i\} \qquad (7.57)$$

This process very closely resembles global stress smoothing sometimes used in the finite element method, Hinton and Campbell (1974). Since MM is positive definite and has a strongly dominant diagonal, its inversion does not present any problems. In short, the global relationship between nodal displacement gradients and displacements can be written as

$$\{V_{i,\alpha}\} = [MNM]\,\{V_i\} \qquad (7.58)$$

where

$$[MNM]_{6m\times 3m} = [MM^{-1}][NM]$$

The numerical solution of the traction BIE, (eq. 7.17), uses the collocation method to form the equivalent system of algebraic equations for the unknowns. Since there are 3n displacement unknowns at n nodes, we take these nodal locations as collocation points.

The continuous representation of displacement gradients given by eq. (7.44) allows for easy and natural treatment of the principal value integrals appearing in eq. (7.17). It can be easily proven that the basic component of all integrals in eq. (7.17) vanishes as the radius of an exclusion circle Γ_ε around the source point goes down to zero, viz.,

$$I_{i\alpha\beta}(P) = \lim_{\varepsilon\to 0} \int_{\Gamma_\varepsilon} V_{i,\alpha}(Q) \frac{r,_\beta(P,Q)}{r^2(P,Q)} dS(Q) = 0 \qquad \begin{matrix} i=1,3 \\ \alpha=1,2 \\ \beta=1,2 \end{matrix} \qquad (7.59)$$

Similarly, on any rectangular patch of elements, symmetrical with respect to the source point at its center (Fig. 7.3), the integral will vanish identically. Any nonsymmetry of the element patch, with respect to a source point, however, results in a rapid growth of this term. The numerical technique employed for the evaluation of integrals resulting from discretization of eq. (7.17) by the interpolation scheme, eq. (7.44), uses the regularization approach

$$\int_{\Gamma_j} \frac{M^j(Q) r,_\alpha(P,Q)}{r^2(P,Q)} dS(Q) = \int_{\Gamma_j} \frac{M^j(Q) - M^j(P)}{r^2(P,Q)} \cdot r,_\alpha(P,Q) dS(Q) \qquad \begin{matrix} j=1,n \\ \alpha=1,2 \end{matrix}$$

$$+ M^j(P) \int_{\Gamma_j} \frac{r,_\alpha(P,Q)}{r^2(P,Q)} dS(Q) \qquad (7.60)$$

where

Γ_j = patch of elements containing node j,

P = source point, located at node j.

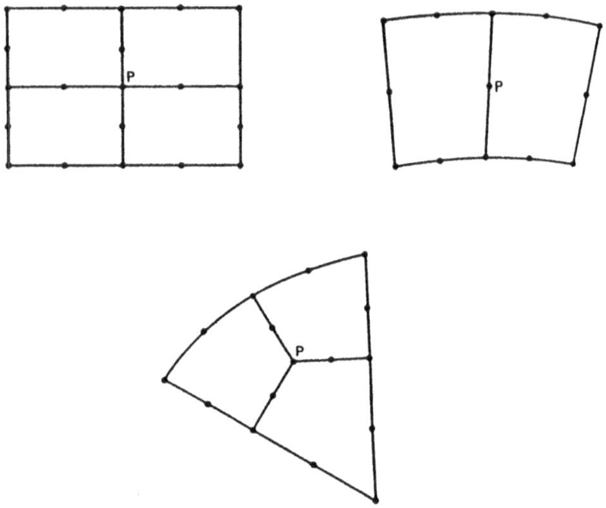

Figure 7.3. Element patches for evaluation of singular integrals

This approach allows for the decrease in the order of singularity of the integrand from $(1/r^2)$ to $(1/r)$ by virtue of the presence of $M^j(Q) - M^j(P)$ term, which is of the order $O(r)$. The first integral of the right-hand side of (7.60), having $(1/r)$ singularity, can now be integrated using the polar coordinate transformation of Rizzo and Shippy (1977). The second integral of the right-hand side of (7.60), containing the nonintegrable (in an ordinary sense) singularity, is handled by the following conversion to a line integral

$$\int_{\Gamma_j} \frac{r_{,\alpha}}{r^2} dS = - \int_{\Gamma_j} \left(\frac{1}{r}\right)_{,\alpha} dS = \begin{cases} - \int_{\partial\Gamma_j} \frac{dy}{r} & \alpha=1 \\ + \int_{\partial\Gamma_j} \frac{dx}{r} & \alpha=2 \end{cases} \tag{7.61}$$

where $\partial\Gamma_i$ is the boundary of Γ_j.

Both terms on the right-hand side of (7.60) can now be handled numerically. A high order of integration is necessary, though, for both surface and line integrals when the source point is located off-center of the patch Γ_j.

The case when a source point lies outside the patch Γ_j of integration is less critical in the sense that the integrals involved are not singular. When the source point is close to the element integrated upon, on the order of shortest element side length away, a very high order of numerical integration is necessary to preserve accuracy. Alternative schemes, like element subdivision, may also be used in this case. In the current numerical implementation, a regular Gauss integration procedure is used with orders of up to 12. Clearly, an improved numerical integration algorithm is needed.

Discretization of the set of eqs. (7.14)-(7.16) with the interpolation scheme of eq. (7.42) at n collocation points (nodes) gives the following system of equations

$$[VV]\,\{V_{i,\alpha}\} = \{\sigma_{\ell 3}\} \qquad \begin{matrix} i=1,3 \\ \alpha=1,2 \\ \ell=1,3 \end{matrix} \tag{7.62}$$

where

[VV] = 3nx6n array of coefficients

$\{V_{i,\alpha}\}$ = 6nx1 column vector of unknown displacement gradients

$\{\sigma_{\ell 3}\}$ = 3nx1 column vector of known surface stresses (tensions)

This underdetermined system cannot be solved directly for the displacement gradients. To reduce the number of unknowns, we use relation (7.58), thus changing the unknowns to displacements. We obtain

$$[VV]\,[MNM]\,\{V_i\} = \{\sigma_{\ell 3}\} \qquad \begin{matrix} \ell = 1,3 \\ i = 1,3 \end{matrix} \tag{7.63}$$

or

$$[VK]\{V_i\} = \{\sigma_{\ell 3}\} \qquad \begin{matrix} \ell = 1,3 \\ i = 1,3 \end{matrix} \tag{7.64}$$

where

$$[VK]_{3nx3n} = [VV][MNM]$$

is the final array of coefficients. Equation (7.64) has 3n displacement components as unknowns and 3n equations formulated at n nodes of the mesh. It can be solved using any convenient linear equation solver after application of displacement boundary conditions at the crack front

$$V_i^{\,k} = 0 \qquad (i=1,3;\ k=\text{boundary node numbers}) \tag{7.65}$$

The structure of eq. (7.64) shows that, for the current formulation for planer cracks, the problem may be separated into two systems of equations. Crack opening displacements are not coupled through the equations with in-plane displacements; thus, the system effectively splits into two systems of equations

$$[VK_{xy}]\{V_i\} = \{\sigma_{\ell 3}\} \qquad \begin{matrix} i=1,2 \\ \ell=1,2 \end{matrix} \tag{7.66}$$

$$[VK_z] \cdot \{V_3\} = \{\sigma_{33}\} \tag{7.67}$$

This property allows for greater efficiency of numerical solution-systems of eqs. (7.66) to (7.67), for the separate subproblems are smaller than the full system. The decoupling of the system is used in the computer implementation of the method. After solution of eqs. (7.66) and (7.67), the displacement gradients at nodes can be obtained using eq. (7.58).

Stresses at any location of the infinite body due to crack loading, except for the crack surface itself, can then be obtained using eqs. (7.14) to (7.16)

$$\sigma_{i\alpha}(p) = - \int_{\Gamma} S_{ji\alpha}(p,Q)V_j(Q)dS(Q) \tag{7.68}$$

Eq. (7.68) presents no particular numerical problem, since the integral is proper. Higher order numerical integration is necessary only when the source point p is located close to the crack surface.

Stress intensity factors at nodes of crack front are calculated using the usual two-dimensional fracture mechanics formulas for plane strain

$$K_I = \frac{V_I E}{8(1-\nu^2)} \sqrt{\left(\frac{2\pi}{d}\right)} \tag{7.69}$$

$$K_{II} = \frac{V_{II} E}{8(1-\nu^2)} \sqrt{\left(\frac{2\pi}{d}\right)} \tag{7.70}$$

$$K_{III} = \frac{V_{III} E}{4(1-\nu)} \sqrt{\left(\frac{2\pi}{d}\right)} \tag{7.71}$$

where V_I is an opening displacement (V_3), V_{II} is an in-plane displacement component normal to the crack front, and V_{III} is an in-plane displacement component tangent to the crack front. The above displacements are for the nodes closest to the crack front, their distance to the crack being d.

7.6 Numerical Results

The computational algorithm presented in the preceding section is composed of two independent parts: the differentiation operator (calculation of nodal values of displacement gradients from nodal displacements) and the traction BIE discretization with respect to nodal displacement gradients. Previous experience [Polch, et al. (1985)] showed that the accuracy of the differentiation operator, eq. (7.59), is crucial to the accuracy of the overall method.

The test of accuracy of differentiation operator consisted of imposing a nodal displacement pattern corresponding to the solution of a buried crack problem under constant pressure on a grid in Fig. 7.4. The resulting pattern of nodal displacement gradients was compared with exact displacement gradients.

Figure 7.5 shows the predicted crack opening displacement gradient relative to the exact solution. Some irregularities were noted at the interface between the regular mesh and the crack front mesh. The midside nodes of the second and third row of elements behind the crack front (see Fig. 7.6) were relocated to positions that created an implied crack opening gradient singularity at the actual crack front. The use of these transition elements smoothed the numerical results.

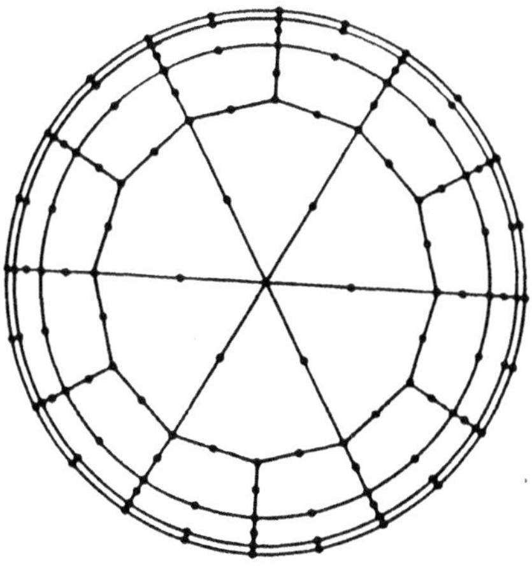

Figure 7.4. Circular crack model

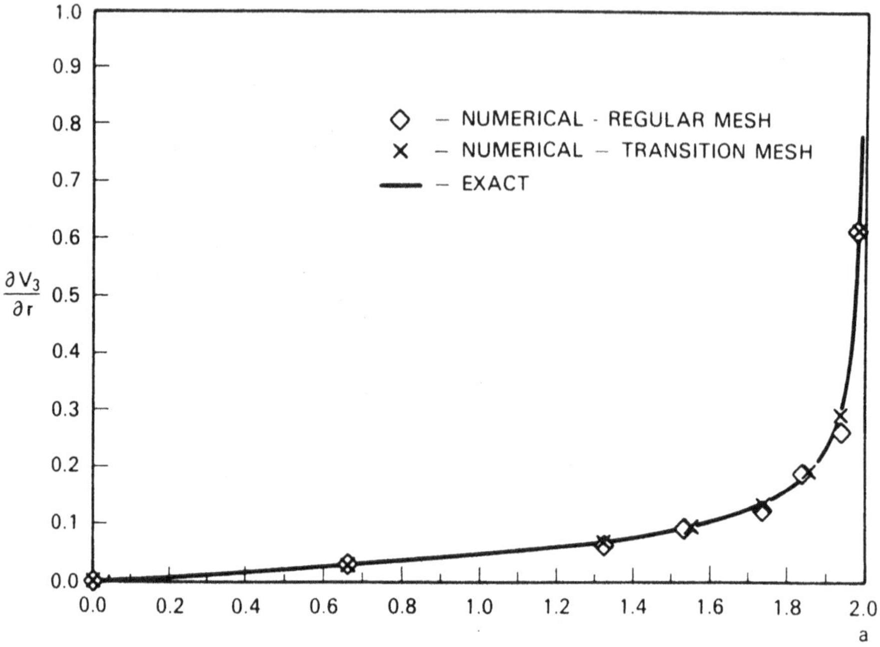

Figure 7.5. Radial distribution of displacement gradients

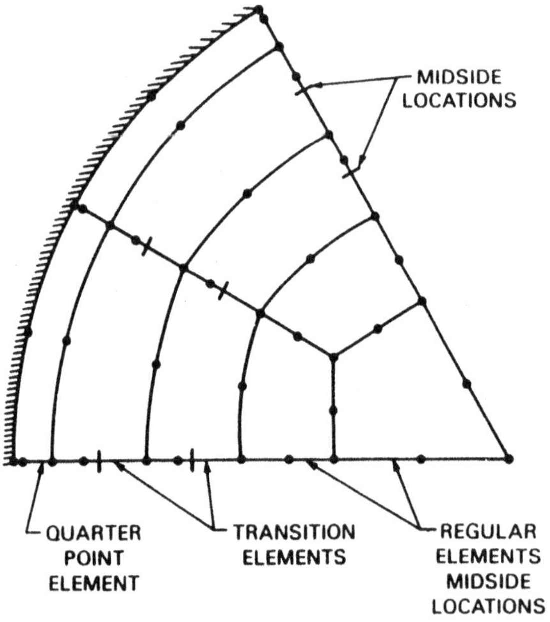

Figure 7.6. Transition element model

The second part of the algorithm, the traction BIE discretization, was checked in a similar fashion to the differentiation operator. A set of nodal displacement gradients corresponding to the exact solution of a buried crack under constant pressure was applied to the discretized traction BIE (7.64). The resulting traction pattern was constant within 9% for the standard mesh and within 5% for the transition mesh as seen in Fig. 7.7. Again, the application of transition elements brought about significant improvement of results.

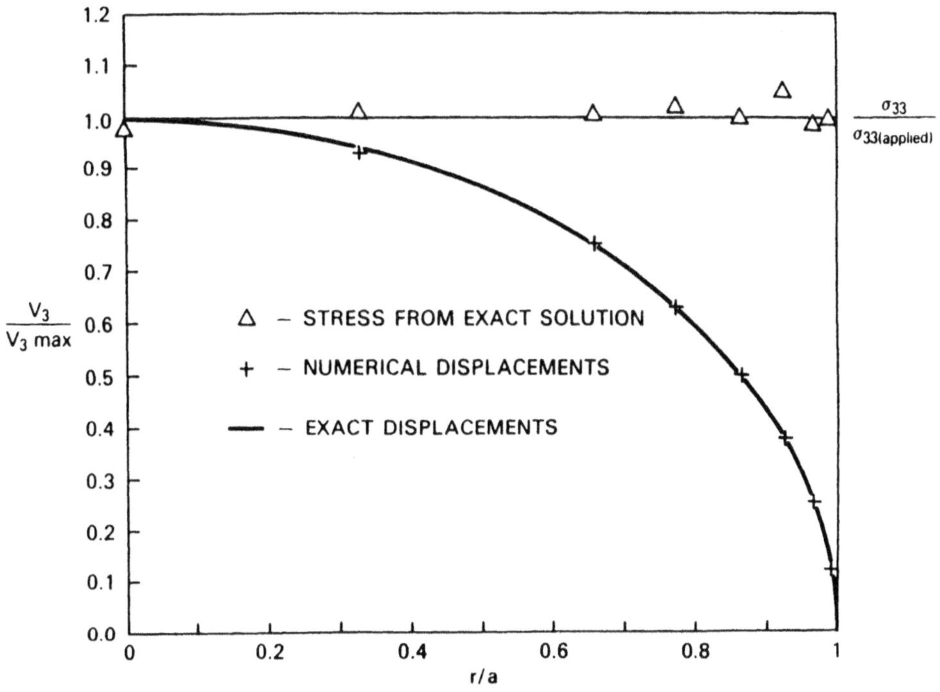

Figure 7.7. Results of computations for circular crack under constant pressure

The full combined algorithm for solution of the buried crack problem was applied to circular and elliptical cracks, as in Fig. 7.8. The circular crack was analyzed with four kinds of loadings: constant pressure, bending moment, and V-shaped and roof-shaped pressure distributions. The loadings were selected in order to compare the results of the method with the exact solutions. The elliptical (a/b=4) crack was analyzed with constant pressure load only. The results of the constant pressure case in terms of crack opening displacement are presented in Fig. 7.7. The error of nodal displacements (and, hence the stress intensity factors from eq. (7.69) is less than 2%.

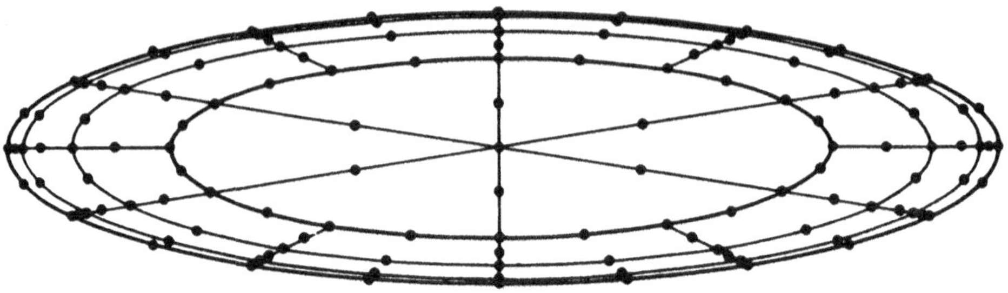

Figure 7.8. Elliptical crack model

The results for the other loading cases are given in Table 7.1, in terms of stress intensity factors at the crack front, as well as the center point crack opening displacement. These results are slightly inferior to those of constant pressure case. Only at the points of abrupt loading change do the results degrade noticeably. Otherwise, the accuracy of results is good.

Table 7.1. Circular crack under various loadings

	Moment Loading ($\sigma_{33} = P(\frac{X}{a})$)			V Shaped Loading ($\sigma_{33} = P\frac{\lvert X\rvert}{a}$)			Roof-shaped Loading ($\sigma_{33} = P(1 - \frac{\lvert X\rvert}{a})$)		
	Numerical	Exact	Error	Numerical	Exact	Error	Numerical	Exact	Error
$\delta\|_{X=0}$	0	0	0	0.0308	0.0393	-21.6%	0.1274	0.1178	+8.1%
α		$K_I/P\sqrt{\pi a}$			$K_I/P\sqrt{\pi a}$			$K_I/P\sqrt{\pi a}$	
0°	0.411	0.424	-3.1%	0.438	0.458	-4.4%	0.185	0.179	+3.4%
30°	0.354	0.365	-3.0%	0.392	0.404	-3.0%	0.232	0.233	-0.4%
60°	0.206	0.211	-2.4%	0.254	0.261	-2.7%	0.369	0.375	-1.6%
90°	0.000	0	0	0.085	0.106	-19.8%	0.539	0.531	+1.5%

References

(1971) R.P. Kanwal, Linear Integral Equations, Academic Press, New York.

(1974) E. Hinton and J.S. Campbell, Local and Global Smoothing of Discontinuous Finite Element Functions Using a Least Squares Method. International Journal of Numerical Methods in Engineering, 8, 461-480.

(1975) R.D. Henshell and K.G. Shaw, Crack Tip Elements are Unnecessary. International Journal of Numerical Methods in Engineering, 9, 495-509.

(1977) F.J. Rizzo and D.J. Shippy, An Advanced Boundary Integral Equation Method for Three-Dimensional Thermoelasticity, International Journal of Numerical Methods in Engineering, 11, 1753 -1768.

(1985) E.Z. Polch, T.A. Cruse, and C.-J. Huang, Buried Crack Analysis with An Advanced Traction BIE Algorithm, in Advanced Topics in Boundary Element Analysis, eds. T.A. Cruse, A.B. Pifko, and H. Armen, AMD-Vol. 72, American Society of Mechanical Engineers, New York.

8. Two-Dimensional Weight Function Evaluation

8.1 Introduction

The weight function method is based on Rice's (1972) interpretation of Bueckner's (1971) original paper. The weight function for a crack problem is generally taken to be the normalized rate of change of surface displacements with respect to crack size for a reference state of loading. As shown by Rice (1972), this weight function acts as a Green's function for the crack problem. That is, the solution to any fracture mechanics problem for the same geometry but different loading conditions can be obtained from the weight function for the reference set of loading conditions. The process involves an integration of the uncracked stress field times the weight function to arrive at the crack tip stress intensity factor for those imposed stresses.

The singular advantage of the weight function method is efficiency of computation of the crack tip stress intensity factor for a variety of crack sizes and loading conditions. Crack size effects such as finite width effects on stress intensity factor, or size effects where crack size changes the applied loads (stiffness effects), need to be included in the weight function. The method reported herein addresses the computational problem of generating weight functions in a direct and efficient manner, while providing a general method for the mixed-mode, mixed-boundary value problem.

The weight function method discussed herein excludes the related, Green's function method for stress intensity factor evaluation. In the Green's function method, many of which are given in Rooke and Cartwright (1976), the stress intensity factor for the reference problem is given in generalized terms as the response to a point load on the crack surface. The point load solution is then integrated as a weighted integral of the applied tractions.

While the Green's function and weight function methodologies are closely related, the computational approaches for the two methods are distinct. The current approach focuses on the numerical evaluation of surface displacement derivatives and tractions to establish the weight function. The approach is based on the boundary element method for two-dimensional fracture mechanics, given in Chapter 5.

Most numerical methods for the development of weight functions are based on boundary integral equation [Besuner (1977)] or finite element methods [Parks and Kamenetzky (1979)]. The boundary integral equation solutions have been based upon numerical differentiation of the numerical results as in Chapter 4. That is, the crack surface displacement numerical solutions are obtained for two slightly different crack lengths. The finite element method was modified by Parks (1979) to include in the virtual work principle an explicit derivative with respect to crack size. This permitted the FEM-based approach to be more efficient than numerical differentiation. This approach has been exploited to a great extent by Sha and Yang (1985).

A second approach to the numerical problem is that proposed originally by Paris, McMeeking, and Tada (1976). In this method, the weight function is computed for the problem of cracked body subject to the elastic singularity tractions on a small circular path surrounding the crack tip. While most of the results were obtained using the FEM, Cartwright and Rooke (1985) used the BEM to obtain very satisfying numerical results. Others, e.g., Grandt (1975) and Petroski and Achenbach (1978), have developed very efficient means of estimating weight functions for a limited number of specific geometries.

8.2 Formulation of the Weight Function BIE

A general purpose numerical evaluation of crack surface weight functions has been developed based on the use of the Green's function boundary element formulation of the two-dimensional fracture mechanics problem. The following sections will present a brief definition of crack surface weight functions and their use. The basic references for these discussions are Rice (1972) for the weight functions and Cruse (1978) for the boundary element method. The general weight function approach of Bortman and Banks-Sills (1983) will be followed.

Consider two loading states for a specified geometry. Applying the reciprocal work theorem

$$\int_{S+\Gamma} t_i^1 u_i^2 dS + \int_{S+\Gamma} t_i^2 u_i^1 dS \tag{8.1}$$

where u_i and t_i are displacements and tractions, respectively, S is the surface of the body excluding the crack, and Γ is the crack surface.

Consider the virtual extension (Δa) of the crack in loading state 2. A singular traction field, t_i^c, with the usual inverse-square root behavior, is taken as state (1) to close the extended part of the crack. The tractions of state (2) ahead of the crack are released, thereby creating the square-root displacement field u_i^c. The reciprocal work theorem applied to modified states yields (assuming no changes in applied loading),

$$\int_{S+\Gamma+\Delta\Gamma} t_i^1 \Delta u_i^2 dS = \int_{S+\Gamma} \Delta t_i^2 u_i^1 dS \tag{8.2}$$

where Δu_i and Δt_i indicate the changes in the displacement and traction fields, and $\Delta\Gamma$ is the change in crack length Γ. Since the integral over $\Delta\Gamma$ on the right-hand side of eq. (8.2) is zero, letting the crack surface to be stress free, we obtain

$$\int_S t_i^1 \Delta u_i^2 dS + \int_{\Delta\Gamma} t_i^c u_i^c d\Gamma = \int_S \Delta t_i^2 u_i^1 dS \tag{8.3}$$

Using the well-known expansions for the crack-tip fields, t_i^c and u_i^c, and taking both crack tips to have identical fields, we obtain

$$\int_{\Delta\Gamma} t_i^c u_i^c \, d\Gamma = -\frac{4\Delta a}{H} \left(K_I^1 K_I^2 + K_{II}^1 K_{II}^2\right) \tag{8.4}$$

where $H = E/(1-\nu^2)$ for plane strain, $H=E$ for plane stress, and K_I and K_{II} are modes I and II stress intensity factors, respectively.

By using the relationship (8.4), eq. (8.3) can be reduced to the required relationship, at the limit $\Delta a=0$

$$\frac{4}{H} \left(K_I^1 K_I^2 + K_{II}^1 K_{II}^2\right) = \int_S t_i^1 \frac{\partial u_i^2}{\partial a} dS - \int_S u_i^1 \frac{\partial t_i^2}{\partial a} dS \tag{8.5}$$

Assuming that the boundary conditions are specified over the same section of the body for both load cases, we have

$$\frac{4}{H} \left(K_I^1 K_I^2 + K_{II}^1 K_{II}^2\right) = \int_{S_t} t_i^1 \frac{\partial u_i^2}{\partial a} dS - \int_{S_u} u_i^1 \frac{\partial t_i^2}{\partial a} dS \tag{8.6}$$

The boundary has been divided into the part for specified tractions (S_t) and that for specified displacements (S_u). Since state 2 is general, we can compute eq. (8.6) from another general independent solution (state - 3). The reciprocal energy release rate is then given by

$$\frac{4}{H} \left(K_I^1 K_I^3 + K_{II}^1 K_{II}^3\right) = \int_{S_t} t_i^1 \frac{\partial u_i^3}{\partial a} dS - \int_{S_u} u_i^1 \frac{\partial t_i^3}{\partial a} dS \tag{8.7}$$

Combining eqs. (8.6) and (8.7), we can solve for the stress intensity factors for the desired problem (state 1) in terms of the solution of the reference problems (states 2 and 3).

$$K_I^1 = \frac{H}{4K} [K_{II}^3 \{\int_{S_t} t_i^1 \frac{\partial u_i^2}{\partial a} dS - \int_{S_u} \frac{\partial t_i^2}{\partial a} u_i^1 dS\} - K_{II}^2 \{\int_{S_t} t_i^1 \frac{\partial u_i^3}{\partial a} dS - \int_{S_u} \frac{\partial t_i^3}{\partial a} u_i^1 dS\}] \tag{8.8}$$

$$K_{II}^1 = \frac{H}{4K} [K_I^2 \{\int_{S_t} t_i^1 \frac{\partial u_i^3}{\partial a} dS - \int_{S_u} \frac{\partial t_i^3}{\partial a} u_i^1 dS\} - K_I^3 \{\int_{S_t} t_i^1 \frac{\partial u_i^2}{\partial a} dS - \int_{S_u} \frac{\partial t_i^2}{\partial a} u_i^1 dS\}] \tag{8.9}$$

where

$$K = K_I^2 K_{II}^3 - K_{II}^2 K_I^3 \neq 0 \tag{8.10}$$

In the case of symmetric loading, eq. (8.8) reduces to

$$K_I^1 = \frac{H}{4K_I^2 K_I^2} (\int_{S_t} t_i^1 \frac{\partial u_i^2}{\partial a} dS - \int_{S_u} \frac{\partial t_i^2}{\partial a} u_i^1 dS) \tag{8.11}$$

This is the same form as developed by Rice (1972), except for the introduction of mixed boundary conditions. The solution approach using weight functions is normally given for $S_u = 0$, such that

$$K_I^1 = \frac{H}{4K_I^2 K_I^2} (\int_S t_i^1 \frac{\partial u_i^2}{\partial a} dS = \int_S t_i^1 h_i^2 dS) \tag{8.12}$$

where $h_i^2(S)$ is the weight function. For the mixed boundary value problem, no single weight function can be written down, and the full form of equation (8.11) has to be used

$$K_I^1 = \int_{S_t} t_i^1 h_i^{2(u)} \, dS + \int_{S_t} u_i^1 h_i^{2(t)} \, dS \tag{8.13}$$

where

$$h_i^{2(u)} = \frac{H}{4K_I^2 K_I^2} \frac{\partial u_i^2}{\partial a}$$

and (8.14)

$$h_i^{2(t)} = \frac{H}{4K_I^2 K_I^2} \frac{\partial t_i^2}{\partial a}$$

Thus, for the general weight function method, one needs an efficient means for solving the reference geometry fracture mechanics problem, subject to any fixed displacement or compliant boundary conditions, to obtain the rate of change in boundary displacements and tractions and the crack tip stress intensity factors. The fracture mechanics BEM presented in Chapter 5 provides the most direct and efficient means for providing this data in the appropriate manner.

The boundary element formulation of the two-dimensional fracture mechanics problem is based on the solution of the following boundary integral equation

$$\mathbf{Cu} + \int_S \mathbf{T}^* \mathbf{u} \, dS = \int_S \mathbf{U}^* \mathbf{t} \, dS \tag{8.15}$$

where the physical variables are the boundary displacement and traction vectors, **u**, **t**. The kernel functions **U***, **T*** in (8) are the displacement and traction (on S) solutions to the elasticity problem for the infinite plane subject to a point force loading.

The weight function method requires the solution in terms of the rate of change of the boundary conditions, as a function of crack length. Differentiation of eq. (8.15) with respect to crack length is completely straightforward due to the explicit dependence on crack length of the kernels of the integral operators. Thus, we obtain the following boundary identity

$$C\,\frac{\partial u}{\partial a} + \int_{S_t} T^{*}\,\frac{\partial u}{\partial a}\,dS - \int_{S_u} U^{*}\,\frac{\partial t}{\partial a}\,dS = -\int_{S}\frac{\partial T^{*}}{\partial a}\,u\,dS + \int_{S}\frac{\partial U^{*}}{\partial a}\,t\,dS \qquad (8.16)$$

The boundary has been denoted as $S = S_u + S_t$ to denote the notions of both displacement and traction boundary conditions on portions of the surface. In general, the mixed boundary conditions involve somewhat more complexity than is denoted by this notation, but the terms in eq. (8.16) convey the essential notion of the algorithm.

Letting the vector $\{\partial \mathbf{x}/\partial a\}$ correspond to the derivative of unknown boundary values (both traction and displacement components), and $\{\mathbf{y}\}$ the known boundary values, a symbolic form of (9) may be written

$$[\mathbf{A}]\begin{Bmatrix}\partial u/\partial a\\ \partial t/\partial a\end{Bmatrix} = [\mathbf{A}]\,\{\partial \mathbf{x}/\partial a\} = [\partial \mathbf{B}/\partial a]\{\mathbf{y}\} - [\partial \mathbf{A}/\partial a]\{\mathbf{x}\} \qquad (8.17)$$

Equation (8.17) is obtained from eq. (8.16) by the imposition of a boundary interpolation system. In the current application, the boundary data is assumed to vary in a piecewise linear fashion, with **x**, **y** representing nodal variables. The algorithm for the solution of eq. (8.17) consists of

1. setting up the discrete form of (8) to obtain **A**,
2. storing **A** for later use,
3. solving for the unknown boundary data,
4. computing the elements of $[\partial \mathbf{B}/\partial a]$ and $[\partial \mathbf{A}/\partial a]$, and
5. solving eq. (8.17) using the entirety of the boundary conditions.

The derivative of the Somigliana identity for the interior displacements corresponding to the boundary solution of eq. (8.16) is given by

$$\partial u/\partial a = -\int_{S} T^{*}\,\frac{\partial u}{\partial a}\,dS + \int_{S} U^{*}\,\frac{\partial t}{\partial a}\,dS - \int_{S}\frac{\partial T^{*}}{\partial a}\,u\,dS + \int_{S}\frac{\partial U^{*}}{\partial a}\,t\,dS \qquad (8.18)$$

Equation (8.18) is equally true for interior points or crack surface points, although care must be exercised in the evaluation of the kernel functions for crack surface points. The numerical solution of eq. (8.18) completes the weight function description, in a formal way. That is, by specifying a suitable number of crack surface locations, the analyst is given a complete description of the rate of boundary condition change with respect to crack length, as needed in eq. (8.11).

However, the term from eq. (8.18) for the point at the crack tip is singular and requires special treatment as discussed in the next section. It should be noted that even though two sets of weight functions are required at the boundary point for the solution of mixed boundary value problems, only one set is required at the crack surface points, where by superposition, a mixed boundary condition problem can be reduced to an uncracked problem subjected to given boundary loadings and to a crack problem with zero displacements specified over S_u section of boundary.

Three example reference problem calculations are presented which illustrate the essential numerical capabilities of the weight function calculation procedure. The first is more in the way of a validation example, as the problem considered is a simple square plate with a central crack, Fig. 8.1. The plate is loaded in simple tension transverse to the crack as shown; this loading results in Mode I response of the crack. The crack sizes range from a/W = 0.01 to 0.5.

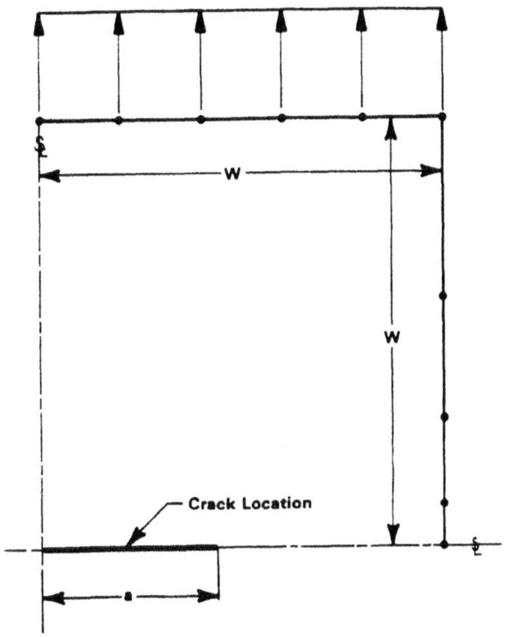

Figure 8.1. Validation example: square plate with central crack

The smallest crack size is essentially equal to the infinite plate problem. The computer code calculated stress intensity factors for each case; for the short crack case the accuracy to the infinite plate result was to five significant figures. Figure 8.2 plots the normalized crack opening displacement derivative results for the four cases. In all cases the resulting distribution is quite smooth over most of the crack; the finite width effect is seen for the cases of a/w > 0.1. The normalized crack tip singular behavior is also seen in Fig. 8.2 to be essentially identical for each of the cases.

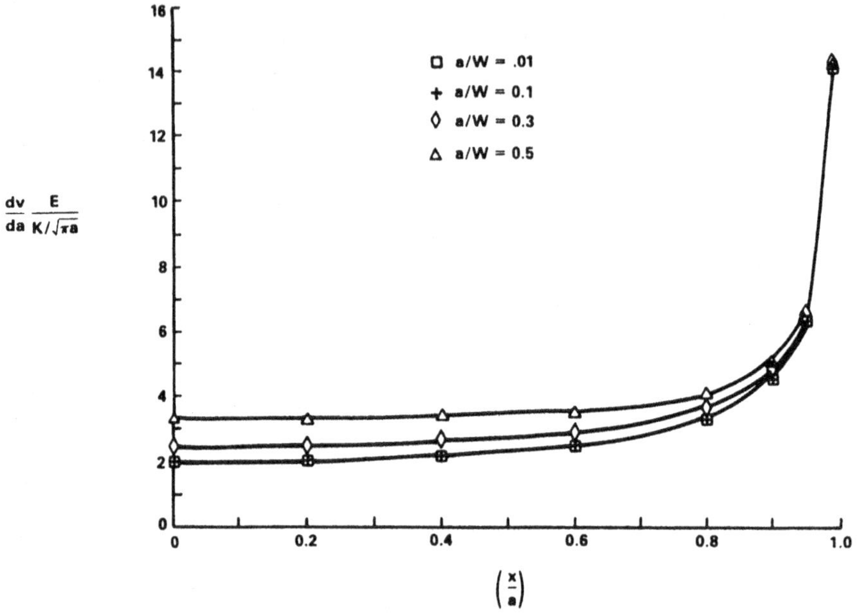

Figure 8.2. Weight function data: tension plate with center crack

Figure 8.3 shows the BEM mesh for the second reference problem considered in this study. The problem is a plate with a central hole and an edge crack from the hole. The mesh was established in a manner that represented mixed boundary conditions, as the left-hand side is taken as a plane of simulated symmetry. The boundary mesh consisted of 27 nodes and was discussed in Chapter 5.

The crack is taken to be located at the horizontal symmetry line, extending from the edge of the crack. It is to be re-emphasized that the BEM algorithm being used does not model the surface of the crack, as this surface is explicitly and exactly accounted for in the formulation of the integral equations. The plate is taken to be sixteen units long, with a width of eight units and a hole radius of two units.

Figure 8.4 plots the computed stress intensity factors for five crack sizes (a/R=1.5, 1, 0.5, 0.25, 0.05). The stress concentration factor for the hole is computed to be 3.59 versus the value of 3.54 from Peterson (1974). The stress intensity factor for an infinitesimally short edge crack at the hold is then given by

$$K_I = 1.122\, K_T\, \sigma \sqrt{\pi a} \qquad (8.19)$$

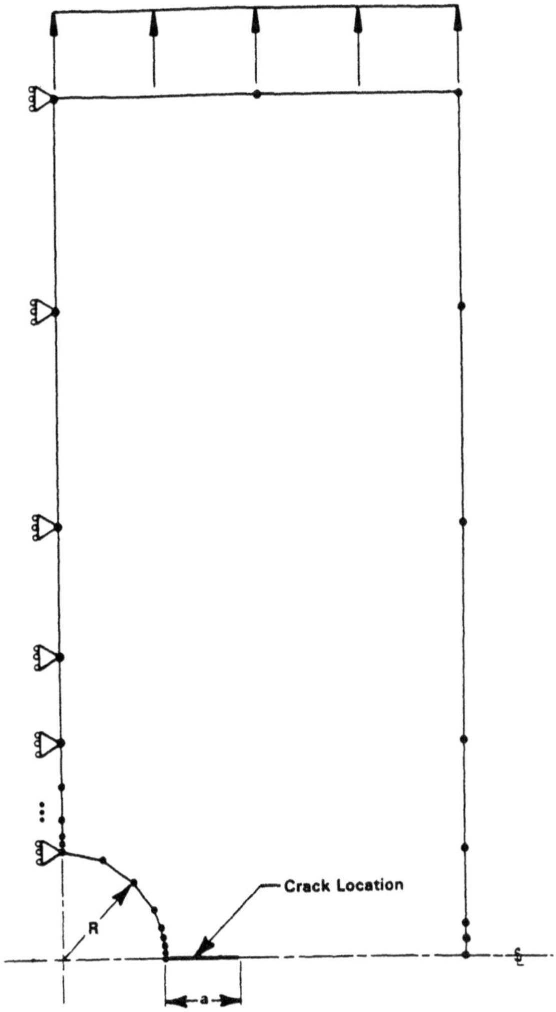

Figure 8.3. BEM mesh for plate with central hole and edge crack

The value of normalized stress intensity factor for A/R=0, therefore, is predicted to be 4.028. The stress intensity factor is seen to decrease with increasing crack size for short crack lengths due to the effect of the stress gradient. For longer crack lengths, the values are seen to begin to rise, as would be expected due to finite width effect. The second curve in Fig. 8.4 simply normalizes the stress intensity factor such that the value of a/R=0 is zero.

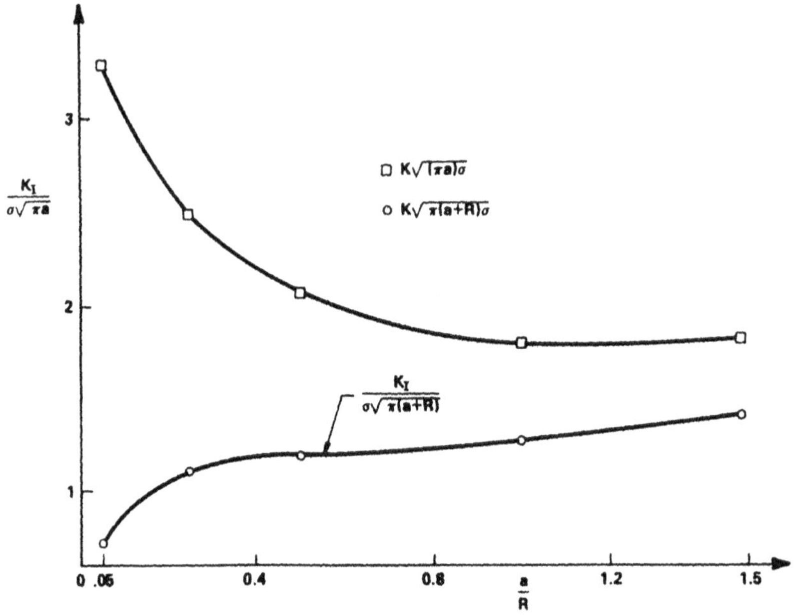

Figure 8.4. Stress intensity factors: tension plate with hole

Figures 8.5 and 8.6 present the numerical results for the normalized weight function of the Mode I reference problem. The results in Fig. 8.5 are essentially of the same shape as in Fig. 8.2, although the zero intercept is elevated due to the free edge effect. Also, the order of the curves is affected by a/R. Figure 8.6 normalizes these results by the numerical results for the center cracked panel (a/W=0). The case for A/R=0.05 clearly shows the influence of the free edge effect for short cracks, while the others begin to look more like the results for a center cracked panel of crack length equal to 2(a+R). The singular behavior of all of these solutions is essentially identical to the center cracked results in the normalized presentation of Figs. 8.2 and 8.5.

The final example selected is a validation example for the mixed-mode case, where a rectangular plate with an inclined crack, Fig. 8.7, is considered. Weight functions computed at crack surface points for two solution states are shown in Fig. 8.8. The stress intensity factor for a third case where the plate is loaded in simple tension was evaluated using the uncracked stresses and the equivalent form of eq. (8.8) at crack surface points. The computed values for normalized stress intensity factors, $K_I/\sigma\sqrt{\pi a}$ and $K_{II}/\sigma\sqrt{\pi a}$, are 0.720 and 0.593, respectively, which are essentially identical to the values directly computed from eq. (5.38).

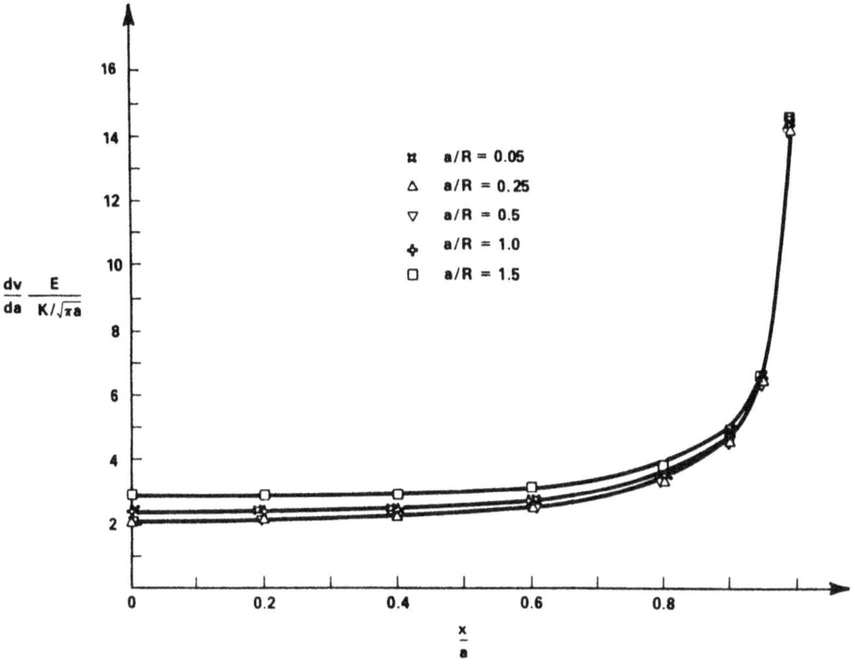

Figure 8.5. Weight function data: tension plate with hole

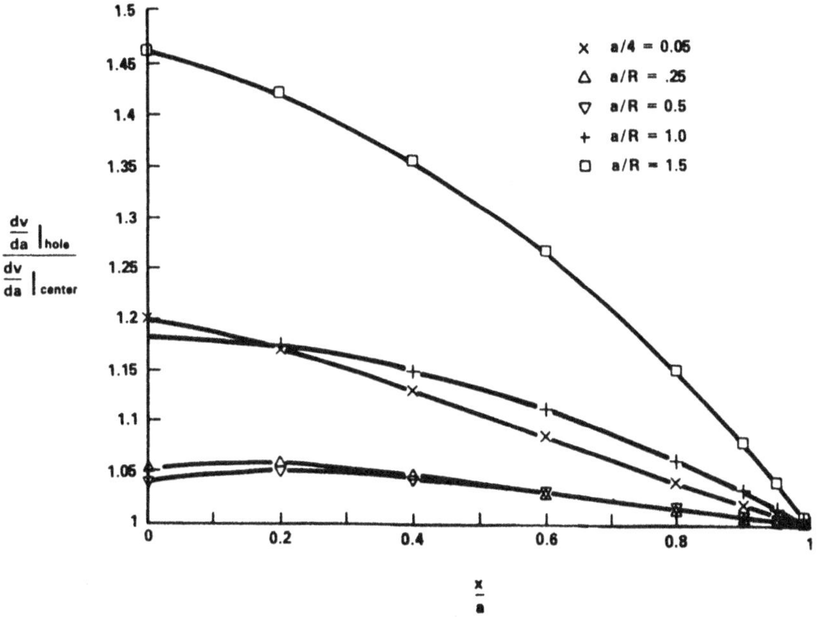

Figure 8.6. Weight function data: tension plate with hole/center crack

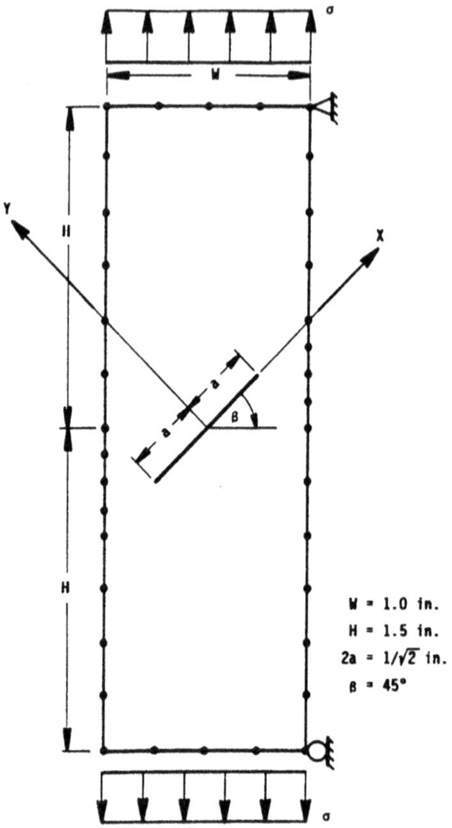

Figure 8.7. Mixed mode example: plate with slant center crack

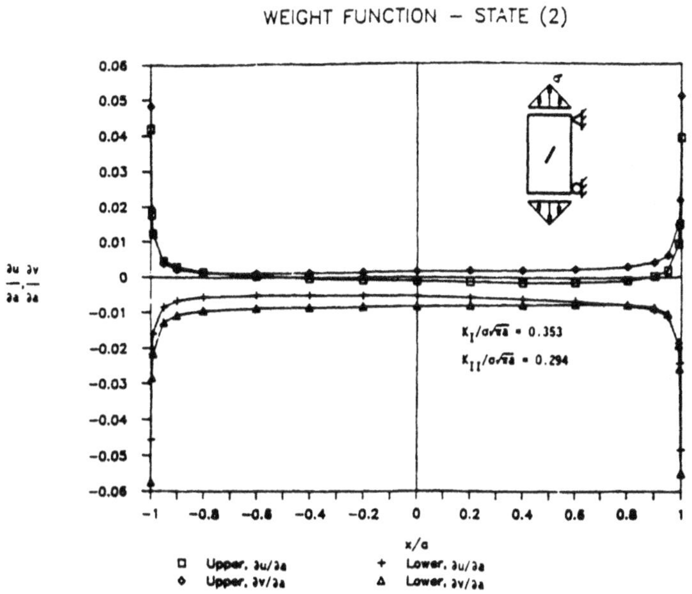

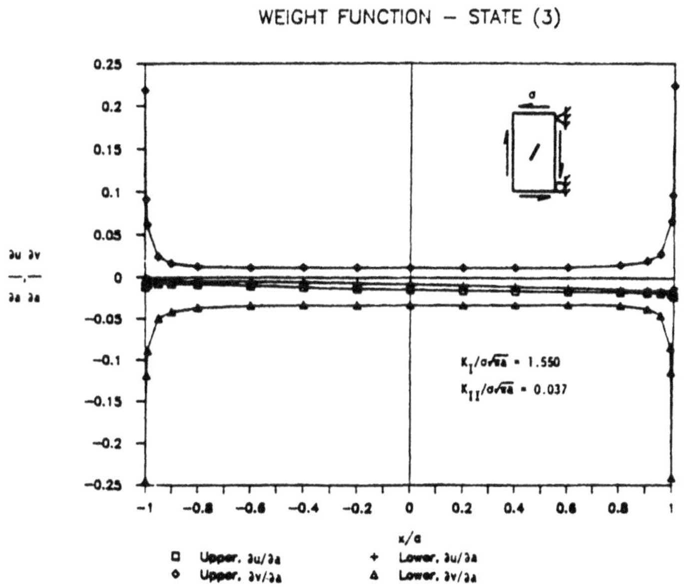

Figure 8.8. Weight function data: plate with slant center crack

References

(1971) H.F. Bueckner, Weight Functions for the Notched Bar, Zitschrift fur Angewandte Mathematik und Mechanik, 51, 97-109.

(1972) J.R. Rice, Some Comments on Elastic Crack-Tip Stress Fields, International Journal of Solids & Structures, 8, 6, 751-758.

(1974) R.E. Peterson, Stress Concentration Factors, J. Wiley & Sons, New York.

(1975) A.F. Grandt, Jr., Stress Intensity Factors for Some Through-Cracked Fastener Holes, International Journal of Fracture, 11, 2, 283-294.

(1976) P.C. Paris, R.M. McMeeking, and H. Tada, The Weight Function Method for Determining Stress Intensity Factors, Cracks and Fracture, ASTM STP 601, American Society for Testing and Materials, 471-489.

(1976) D.P. Rooke and D.J. Cartwright, Compendium of Stress Intensity Factors, Her Majesty's Stationery Office, London.

(1977) P.M. Besuner, The Influence Function Method for Fracture Mechanics and Residual Fatigue Life Analysis of Cracked Components Under Complex Stress Fields, Nuclear Engineering & Design, 43, 115-154.

(1978) T.A. Cruse, Two-Dimensional Boundary-Integral Equation Fracture Mechanics Analysis, Applied Mathematical Modeling, 2, 287-293.

(1978) H.J. Petroski and J.D. Achenbach, Computation of the Weight Function from a Stress Intensity Factor, Engineering Fracture Mechanics, 10, 257-266.

(1979) D.M. Parks and E.M. Kamenetzky, Weight Functions from Virtual Crack Extension, International Journal of Numerical Methods in Engineering, 14, 1693-1706.

(1983) Y. Bortman and L. Banks-Sills, An Extended Weight Function Method for Mixed-Mode Elastic Crack Analysis, Journal of Applied Mechanics, 50, 907-909.

(1985) D.J. Cartwright and D.P. Rooke, An Efficient Boundary Element Model for Calculating Green's Functions in Fracture Mechanics, International Journal of Fracture, 27, R43-R50.

(1985) G.T. Sha and C-T. Yang, Weight Function Calculations for Mixed-Mode Fracture Problems with the Virtual Crack Extension Technique, Engineering Fracture Mechanics, 21, 6, 1119-1149.

(1986) T.A. Cruse and E.Z. Polch, Application of an Elastoplastic Boundary Element Method to Some Fracture Mechanics Problems, Engineering Fracture Mechanics, 23, 439-452.

(1986) T.A. Cruse and E.Z. Polch, Elastoplastic BIE Analysis of Cracked Plates and Related Problems, International Journal for Numerical Analysis in Engineering, 23, 429-437.

INDEX

T

U

V

W

X

Y

Z

GPSR Compliance

The European Union's (EU) General Product Safety Regulation (GPSR) is a set of rules that requires consumer products to be safe and our obligations to ensure this.

If you have any concerns about our products, you can contact us on ProductSafety@springernature.com

In case Publisher is established outside the EU, the EU authorized representative is:

Springer Nature Customer Service Center GmbH
Europaplatz 3
69115 Heidelberg, Germany

Batch number: 10761599

Printed by Printforce, the Netherlands